THE BOOK OF THE AMERICAN WOODCOCK

October—An Etching

There where the woodcock his long bill among the alders
Forward in level flight propels,
Tussocks of faded grass are islands in the pasture swamp
Where the small foot, if it be light as well, can pass
Dry-shod to rising ground.

Not so the boot of the hunter.
Chilly and black and halfway to the knee
Is the thick water there, heavy wading,
Uneven to the step; there the more cautious ones,
Pausing for a moment, break their guns.
There the white setter ticked with black
Sets forth with silky feathers on the bird's track
And wet to his pink skin and half his size comes back.

Cows are pastured there; they have made a path among the alders.
By now the keeper's boy has found
The chalk of the woodcock on the trampled ground.

Edna St. Vincent Millay

FRONTISPIECE AND ETCHINGS BY A. LASSELL RIPLEY

THE BOOK OF THE AMERICAN WOODCOCK

by William G. Sheldon

Published by agreement with the
United States Department of the Interior, Bureau of Sport Fisheries and Wildlife

THE UNIVERSITY OF MASSACHUSETTS PRESS

Color frontispiece courtesy
of the Frederic C. Walcott Memorial Fund of
the North American Wildlife Foundation

The University of Massachusetts Press in Amherst

Library of Congress Catalog Card Number 67-11243
Printed in the United States of America

Second printing, with corrections, 1971

To the memory of William O. Clark

superb sportsman and friend

This volume is a contribution of the Massachusetts Cooperative Wildlife Research Unit supported by the Bureau of Sport Fisheries and Wildlife, the Massachusetts Division of Fisheries and Game, the Wildlife Management Institute and the College of Agriculture (Department of Forestry and Wildlife Management) of the University of Massachusetts.

Preface

The American Woodcock (*Philohela minor*, Gmelin) is a shorebird which has forsaken its ancestral haunts of mud flats and marshes for the uplands. Abroad primarily at night or in the twilight of evening and dawn, this bird has long offered a challenge to sportsmen and scientists, who seek to unravel its life history.

Specialized animals requiring a narrow range of habitat are highly vulnerable to decimation since their evolution has proceeded so far that they are incapable of adapting themselves to a wide variety of environments. When man alters the land, some kinds suffer serious consequences. The woodcock is such a species. With its long bill, which is prehensile at the tip, the bird can pick up invertebrates from the surface of the land, but depends largely on damp ground where it can thrust its bill to the hilt, securing such food as angleworms. The drainage of wetlands irreparably destroys woodcock habitat, and the impact of this activity has been most serious in the vital wintering grounds of Louisiana. When man blankets the land with lethal chemicals, the woodcock's food may become so contaminated as to cause sterility or to kill it outright. When housing developments invade the woodcock's resting places on its traditional migration routes, it suffers stress. As surely as the Ivory-billed Woodpecker (*Campephilus principalis*) depends on quantities of large decaying hardwood trees, or the American Brant (*Branta bernicla hrota*) on the presence of eel grass, the woodcock needs the moist land of forests in early stages of succession.

The woodcock is enough of a recluse to be an inconspicuous species. Many men have lived their lifetimes in areas where woodcock are common without having seen one. Others have flushed this unique bird from the cover beside dimly lit trails without knowing what it was.

The objectives of this study have been to gather as much information as possible on the ecology and life history of the woodcock, bringing under one cover much that is known today about this fine upland game bird. Major phases of the investigation have included (1) Behavior and activities on the breeding grounds and wintering grounds; (2) Migratory habits; (3) Distribution and abundance; (4) Factors affecting the welfare of the species; (5) Management techniques.

I have studied the woodcock in its breeding grounds for fifteen years. Since my research was carried out in Massachusetts and other New England states, it is inevitable that the study is oriented more to the Northeast than to other parts of the woodcock range. However, I have drawn freely on the works of others in the Lake States and in the South. Basically, the needs and behavior of woodcock differ only in details from one area to another. A woodcock is a woodcock whether he breeds near James Bay or in Louisiana. Feeding flights at dusk, for example, occur in all areas at all seasons, varying only in the numbers involved. The vegetation of habitat differs widely from one area to another, but the basic requirements of young growth and other characteristics described in this volume differ little whether the bird is living in the bayous of the South or the mixed forests of the North.

The results of many of my studies have been published in various scientific journals (Sheldon, 1952, 1953a, 1953b, 1954, 1955, 1956a, 1956b, 1956c, 1960, 1961, *et al.* 1965a, *et al.* 1965b). In drawing upon these earlier studies in this volume I have revised the material to bring it up to date. To provide details that will be of interest primarily to professional biologists I have included a liberal appendix.

In preparing this study, my greatest debt of gratitude is to fellow woodcock investigators who have generously turned over to me unpublished results of their own research. Heading the list is Leslie L. Glasgow who has made the most intensive studies of anyone on the woodcock wintering grounds. William H. Marshall contributed material from his Minnesota studies, and Bruce Wright from his New Brunswick research. Some unpublished theses on woodcock were made available by Howard L. Mendall of Maine. His fine earlier publication on woodcock in Maine (Mendall and Aldous, 1942) was freely drawn on. Stephen Liscinsky of Pennsylvania permitted the use of many of his findings. A major contribution was the unpublished study of woodcock in Michigan by L. H. Blankenship. My colleague, Fant W. Martin, who was assigned as a full-time woodcock investigator by the Bureau of Sport Fisheries and Wildlife in 1960, has not only extended our knowledge of woodcock on a national basis, but has been most cooperative in supplying me with his findings, some of which still remain unpublished. Most of these workers have reviewed this study and offered many helpful suggestions. Pettingill's (1936) early monograph on woodcock, which included an extensive library research, was a fruitful source of some of the observations recorded in this volume.

Studies in the wintering grounds included important ones by Norris (1941 and 1942) and a few others of less intensive nature. L. J. Merovka (1939), U. S. Game Management Agent, was among the first investigators to gather data on woodcock abundance in Louisiana and to encourage further study. In 1940 E. A. McIlhenny of Avery Island, Louisiana, reported on the effects of cold weather on woodcock and also on a technique used in capturing woodcock for banding. V. H. Reid and P. D. Goodrum, U. S. Fish and Wildlife Service, made winter censuses in central Louisiana from 1949 to 1956. J. J. Lynch, U. S. Fish and Wildlife Service, reported on the effects of severe weather on woodcock during the winter of 1951–52.

H. A. Hochbaum deserves special credit for reviewing an early draft. Others who have critically read the manuscript are Van Harris of the Bureau of Sport Fisheries and Wildlife, Frederick Greeley, David K. Wetherbee, G. A. Ammann, C. H. D. Clarke, Gustav Swanson, J. A. Hagar, and Robert McClung. Joseph Hickey reviewed the section on population dynamics, and Chandler Robbins, the chapter on breeding habits.

I am especially indebted to personnel of the Wildlife Management Institute. James L. Trefethen spent many hours editing this manuscript in its early stages and deserves great credit. Ira Gabrielson reviewed the manuscript, and C. R. Gutermuth has been most helpful and has encouraged publication from the beginning.

A host of graduate students and outside assistants made this work possible. Ethan V. Howard, Jr. was my righthand field assistant for many years. Others who contributed to the field work include Gardner Hobart, Gordon Hobart, T. D. Ford, Robert Wood, Wesley Jones, Thomas Ripley, W. A. Fitzpatrick, Lewis Cowardin, Douglas Bysiewski, Wendell Dodge, James Baird, Herbert Maxfield, R. Gehling, Leslie Campbell, John Kupa, and many others.

This study was originally initiated through the encouragement of the late Logan J. Bennett whose untimely death took a staunch conservationist from our ranks. He was a recognized authority on woodcock.

I am particularly grateful to A. Lassell Ripley for permission to reproduce several unparalleled woodcock etchings and the watercolor used as frontispiece. I am indebted to A. H. Underhill for loaning the last. Alfred Godin prepared most of the other illustrations, and Eva-Maria Fenner, the balance; they both deserve high credit for their fine work.

Lawrence Bartlett was helpful in working out the molt pattern of woodcock and in sponsoring special anatomical studies of the bird by some of his students.

Among many fellow sportsmen who have helped immeasurably, not only in reading sections of this manuscript but in contributing valuable information from their hunting journals, are Robert Livermore, Richard Borden, George Weed, and the late W. O. Clark.

To my wife, Louise, who tended the home fires on many a late night, and aided on spring and summer trips to count birds, I am deeply grateful.

WILLIAM G. SHELDON

Amherst, 1966

Contents

Preface, ix

Prologue, 1

CHAPTER I The Study, 7

CHAPTER II History, 16

CHAPTER III Range, 21

CHAPTER IV Physical Characteristics, 29

CHAPTER V Breeding Habits, 44

CHAPTER VI Feeding Habits, 78

CHAPTER VII Migration and Homing, 95

CHAPTER VIII Hunting, 114

CHAPTER IX Management, 127

CHAPTER X Population Dynamics, 140

CHAPTER XI European Woodcock, 160

Epilogue, 173

APPENDIX I Trapping and Netting, 177

APPENDIX II Figures and Tables, 185

APPENDIX III List of Plants, 200

APPENDIX IV Key for Aging Woodcock from Wings, 203

Selected Bibliography, 207

Index, 219

Illustrations

1 Flight woodcock, 3
2 Banding woodcock, 9
3 Glasgow's handnet, 11
4, 5 Quabbin Reservation singing fields, 13
6, 7 Range maps, 23
8 Sex differentiation by primaries, 31
9 Cerebral axis, 32
10 Comparative brain positions, 33
11 Distal end of upper bill, 34
12 Skull of woodcock, 36
13 Summer weights of 152 woodcock, 41
14 Mean seasonal weights of 940 woodcock, 41
15 Early woodcock, 45
16 Male approaching decoy, 48
17 Male copulating with decoy, 49
18 Male before copulation, 50
19 Tail fanned in alarm, 51
20 Tree height and size of singing field, 63
21 Early woodcock nest, 67
22 Prime nesting cover, 68
23 Over-mature nesting cover, 69
24 Hatching dates, 1950–1961, 70
25 Incubating hen, 71
26 Freshly hatched chicks, 73
27 Fleeing juvenile woodcock, 74
28 Hen with chicks, 75
29 Decoy flight of hen, 76
30 Woodcock probing, 83
31 Woodcock feeding, 84
32 Summer evening flight patterns, 87
33 Winter feeding field, Louisiana, 91

34 Migration routes, 97
35 After woodcock, 115
36 Hunting cover, Michigan, 117
37 Dog pointing brood, 118
38 Flushing woodcock, 129
39 Alders in lumbered fir and cedar swamp, 131
40 Alder cover, Massachusetts, 135
41 Ideal aspen covert, 137
42 American and European woodcock, 161
43 American and European woodcock wings, 162
44 Setting decoy trap, 178
45, 46 Decoy trap, 180
47, 48 Setting mist nets, 182
49, 50 Singing and summer netting fields, 183
51 Courtship flights of single male, 185
52 Time on ground between flights, 185
53 Duration of courtship performances, 186
54 Average number of peents, 186
55 Mean census counts, 187
56 Sex and age ratios, 187
57 Numbers of wildfowl and woodcock hunters, 187
58 Aging by secondary feather patterns, 203

Tables

1 Dates of heaviest woodcock kill, 106
2 Return to singing grounds, 111
3 A Michigan hunting record, 120
4 Michigan woodcock flushes, 120
5 Estimated woodcock kill, 121
6 Flushes per gun hour, 130
7 Sex ratios of woodcock, 142
8 Sex ratios of galliform birds, 142
9 Woodcock survival, 146
10 Life equation, 147
11 Breeding density, 148
A Molt pattern, 188
B Summer weights, 189
C Changes in summer weights, 189
D Fall weight increase, 189
E Average fall weights by sex, 189
F Average weights by age and sex, 190
G Mean weights by age and sex, 190
H Courtship performances, 190
I Singing periods, 191
J Counts of singing males, 191
K Size of singing fields, 191
L Vegetation on singing grounds, 192
M Hatching dates, 194
N Food items, 194
O Band returns in Massachusetts, 195
P Sex and age ratios for Massachusetts, 196
Q Band returns in Louisiana, 196
R Overstory density, 196
S Age and sex ratios, 197

THE BOOK OF THE AMERICAN WOODCOCK

Prologue

In early May the northern woods were vibrant with the reawakening of life. In a stand of aspens twenty feet high the first warbler arrivals were feeding in the treetops. The muffled thunder of the drumming Ruffed Grouse resounded through the nearby hills. Does, heavy with fawn, were seeking secluded spots for the most important event in their lives, which would take place a month later. Raccoons were nursing their newly born young, and fox pups were already playing in front of their dens. Most birds were staking out territories in preparation for mating, nesting and rearing young.

Woodcock had arrived a month earlier. These birds had abandoned the warmth of the South during the first full moon in early February and arrived on their nesting grounds while snow patches still mottled the earth and sub-freezing temperatures at night skimmed the puddles with ice.

By mid-April one hen had completed her clutch of four brown speckled slate-colored eggs. She had wasted no time in making a nest, scratching out a cup-shaped depression in the dead leaves at the base of an aspen. Twenty one days later she felt the first stirrings of life under her breast feathers, and the male chick, with his brother and two sisters, awoke from his days of incubation and broke free of the shell that imprisoned him.

At dawn the next morning, he took his first steps. With neither ceremony nor warning, his mother left the nest and walked a few yards away. In response to her clucking, the chicks, still weak from the ordeal of hatching, struggled after her in an instinctive effort to regain the shelter of her outspread wings. By the standards of any creature but a woodcock they were grotesque bits of buff-and-brown fluff. Their shoebutton eyes were too big and set too high above enormous bills that continually caught on leaf stems and twigs as they struggled to complete their first journey, and the feet that supported them were nearly as large as their mother's. They never returned to the nest.

By the second day, although they still felt uncertain and weak, the brood followed their mother farther from the nest. She overturned leaves and probed in the damp soil to bring forth choice worms, grubs and other invertebrates which she passed to her chicks. The chicks ate ravenously. By the third day they were learning to forage on their own. The male chick's ungainly bill, growing rapidly each day, soon became more efficient. To satisfy his voracious appetite he fed in the evening, part of the night, and from dawn until two hours or more after daylight; as the days passed, he grew rapidly. His mother led him and his brood mates along small brooks, through damp alder swales, or to damp rich loam under short aspens or birch. By the time he was four weeks old he could fly several hundred yards and was almost fully feathered and grown.

Two weeks later he became more adventuresome than his brother and sisters. Each evening he flew as far as two miles to dry, upland fields where he joined birds of other broods in searching for new foods—beetle and fly larvae, caterpillars, and ants. During his first flights he still uttered the *wheep* note of a chick, but by mid-July he attempted tentative peenting notes such as his father made in the spring.

During the heat of the day he rested in cool shaded areas under dense stands of conifers or in damp thickets. He was strong in flight but had gained little weight since early July, when he weighed approximately four ounces. But by mid-August hard fat began to accumulate on his body.

At that time, too, he began to molt his juvenile plumage. By the time of the first frosts in September, he was clothed in an entirely new set of feathers except for the primaries and secondaries of his wings. By now, he was making evening flights of several miles. Then, one evening, he cut the ties with his home territory and flew a full forty miles to the west. The upland insects on which he had fed in the hot evenings of summer had disappeared in freezing temperatures; he now became especially dependent on his long bill,

Figure 1. Flight woodcock (Etching by A. Lassell Ripley)

which he used chiefly for extracting earthworms from the soil. By this time he weighed about six ounces.

In October, he had a new experience: dogs and men invaded his covert. Once, as he flushed, there were loud explosions behind him, and he felt a sting of pain in his right foot. He had lost one toe, but the injury proved only a minor inconvenience.

It was 10° below freezing the night after this adventure. At dusk, in response to a strong urge, he and other birds remaining in the covert flew higher than usual and turned south. The night was clear, with a full moon and a strong breeze from the northwest. In scattered formation the birds flew nearly fifty miles, alighting before dawn in a large alder swamp. They fed vigorously in seepage areas until daylight. Some stayed in the alders all day, feeding sporadically. Others left at dawn and scattered to upland retreats.

Our bird and his group of migrants had swung southeast on the first leg of their southern journey, stopping near the seaboard. They stayed two or three days and then proceeded south by leisurely hops, with layovers of a few days at most stops. At times they passed over stretches of open ocean, crossing Cape Cod, and stopping for three days on Martha's Vineyard. Subsequently they followed the Atlantic coast, passing through Connecticut, New York, New Jersey, Delaware, Maryland, Virginia, and North Carolina. Our male joined the birds that swung southwest and crossed south of the main range of the Appalachians. He ended his migration on December fifteenth in the great swamps along Bayou Blue in Louisiana.

Here he joined hundreds of other woodcock from the great expanse of the northern breeding grounds. Each evening at dusk he flew to pastures or cornfields to feed with other birds on earthworms. He usually ate from sunset until about 11:00 P.M., rested until an hour before dawn, then ate again until early daylight, when he returned to his daytime retreat.

As the days lengthened in late January, the woodcock began short courtship flights. Leaping upward he would rise in spirals, with twittering wings, to a height of three hundred feet. Then descending like a falling leaf, he would utter a series of liquid chirps. Landing on the ground he would give voice to intermittent *peents*, each preceded by a soft gurgling *tuko*.

He left Louisiana with other males in late January. The larger females would follow them later. Stimulated by the mating urge, the yearling males crowded close behind the margin of the retreating snow in their eagerness to return to the northland. In Connecticut, a late snowstorm, followed by a heavy freeze, caught them for nearly a week, and some of the birds died.

He reached his breeding ground in Maine in early April. But older, more experienced males had arrived before him. When he started his search for a suitable courtship territory, the earlier claimants pursued him, cackling defiantly. Finally, after a dozen such rebuffs, he found a small grassy field studded with scattered alders, meadowsweet, and young pines and birches. It was less than a quarter of a mile from the swale where he himself had hatched the year before. No other male had claimed it.

Here, after resting from the exertions of his flight, he began to sing in an effort to attract the females, who were beginning to arrive from the south. Each evening at dusk and each morning at dawn, he spiraled skyward, uttering the breeding call of his kind. On the third evening, his efforts were rewarded when a female fluttered down beside him. He mated with her, but this was the extent of his parental duties. While she sought seclusion in which to nest and raise their offspring, he continued his breeding flight and song. In the next few weeks, before mid-June, he had mated with other females. Then as the breeding urge left him, his annual life cycle completed, he spent the summer in a celibate search for food, rebuilding his energy for his long southern flight in the fall.

He completed two more round trip migrations, returning each year either to the same singing field or one closely adjacent, constantly challenged by young males who were seeking breeding sites of their own. During a snowstorm on his spring migration to the north in his fourth year he met his death by colliding with a power line crossing his migration route. That spring a younger male took over his singing field.

This typical life pattern evolved over thousands of generations of woodcock. It can continue only so long as we reserve some living space for this ghostlike bird of the uplands. Endowed with fortitude and a certain nobility, the bizarre little American Woodcock excites the wonder of ornithologists and elicits the highest ethics of sportsmen. It is a natural resource of rare quality, offering recreation and esthetic pleasure to thousands, and one we should forever be vigilant to preserve.

CHAPTER I

The Study

FIELD METHODS

Spring Captures

On the evening of April 3, 1949, I heard the first woodcock of this study on his spiral courtship flight over my house in Shutesbury, Massachusetts. I found his courting field the next evening at the site of an old sawmill 200 yards behind my house. As I watched him on subsequent evenings I recalled the manually operated trap baited with a stuffed female woodcock that Norris, Beule, and Studholme (1940) had used to catch 23 males on singing grounds in Pennsylvania.

I fashioned an automatic trap out of heavy wire and netting on the same principle as a steel trap but with a pan 12 inches square and a jaw spread of 4 feet shaped like a large butterfly; it was baited with a stuffed female. It was an unwieldy monstrosity, but it worked. I trapped the first male woodcock I tried it on. Unfortunately it caught only four more birds that spring.

In the following summer I developed a far more efficient trap on the same principle (Figures 45, 46, Appendix I). Traps of this pattern were used for the next six years to catch 795 male woodcock and 17 females in spring. During this time I gained additional experience and refined my trapping technique (Sheldon 1955, Appendix).

William Gunn (1951), a Canadian wildlife biologist, had made a tape recording of a male woodcock on its singing ground. He visited Amherst in the spring of 1950; playing the tape elicited a defensive response from a resident bird. Using this recording and nets, Gunn trapped several birds in Canada. I had a record cut from this tape and rigged 12 portable victrolas with speakers. These were of limited success, and required an operator at any site where they were used, whereas one man could operate 12 singing field traps.

I experimented with mist nets in May of 1964. The technique proved so successful that I recommend it as the most efficient and productive method of capturing woodcock for banding or study in the spring. Nets are available in different sizes and weights. The mesh is strung on heavy horizontal threads which are taut. There is sufficient slack between these lines so that a bird hitting the mesh becomes entangled (Appendix I). I set two 30-meter four-tier Japanese thrush nets on eight-foot poles at right angles to each other close to the male's alighting spot on a singing field. The nets were set at several sites the same evening and tended after singing ceased. Defining a trap night as two nets set at one site, I caught 28 woodcock in 27 trap nights as compared to one male in four trap nights with a decoy trap. Two woodcock were trapped the same night at six sites. Three birds hit nets at one site the same night, but one escaped. Two of these "extra" birds were females. I failed to catch birds at five sites; in two instances, high winds disarranged the net, and three sets of nets were erected at sites where the alighting spots were not exactly located. The netting took place between May 5 and May 13, during the height of the woodcock hatch. I believe that if such netting were begun late in March and early April, many females would be captured. With nets properly placed, success should approach 100 per cent. Eighty-five per cent of the captured birds were caught in the bottom two tiers. No birds were netted in the top tier. There are minor disadvantages. Sometimes birds are slightly injured if badly entangled in the net. In May, nets should be erected as late in the evening as possible to avoid the nuisance of extracting passerine birds, which are abundant in woodcock coverts.

Ammann (1963), using a modification of a Bobwhite Quail trap (funnel principle), was successful in capturing a few adult hens, luring them in by the call of captured chicks enclosed in the trap. Mendall and Aldous used a bird dog to find broods for banding. Wright (1960, personal communication) in New Brunswick and Liscinsky (1960, personal communication) in Pennsylvania also

Figure 2. Banding woodcock (Photo by L. A. Campbell)

used dogs for this purpose. My English setter found 28 broods and 3 nests. I have found dogs helpful in all seasons; they locate birds, broods, or nests that I could not find by other means, hence they are valuable aids in defining habitat.

Summer Captures

During the first three years of my study I had difficulty finding woodcock in summer. In late June, 1952, however, I found birds crossing a section of dirt road in my study area from west to east each evening at dusk. I stationed six graduate students on the east side of the road on the evening of July 8, 1952. The evening was humid with a temperature of 80° F. Woodcock started crossing the road at 8:00 P.M. half an hour after sundown and just after the first whippoorwill began to sing; 44 were counted in the next 15 minutes. They disappeared over a ridge to the east.

Several of us followed the line of flight for the next ten evenings; most birds lit in a small opening of less than an acre. Later we found three additional fields that woodcock were using.

Japanese mist nets were strung up on poles 20 feet high in three of these fields and 773 woodcock were captured between late June and September from 1952 to 1960 (Sheldon, 1960). Erecting 20-foot poles was not only time consuming and laborious work, but the high structure drove most of the birds from the field after a few days. Because of these handicaps I recommend using ten-foot poles with a four-tier net for all summer netting. Unlike the singing ground traps, the nets are indiscriminate and birds of both sexes and all ages are caught.

Although I found the mist net most efficient, there are other methods of capturing woodcock in summer. Before I developed the netting technique I caught 15 birds in summer by setting the spring singing ground trap without a decoy in feeding areas. Twenty-two birds were caught in funnel traps set primarily for Ruffed Grouse. Funnel traps with 50-foot lead fences a foot high in feeding areas were used by Liscinsky and Bailey (1955) in Pennsylvania to trap 61 woodcock in 1952 and 1953.

Dr. Fant Martin (1964) and Eldon Clark (1965) captured 605 birds by mist nets, 531 by funnel traps and 361 by lights and hand nets during the summers of 1962 to 1965 in Maine and Michigan. Mayer (1828) captured European Woodcock (*Scolopax rusticola*) with "draw" nets set at dusk across air routes used by the birds. Mouseley (1935), who made an historical study on the European bird, reproduces a woodcut made before 1629 of netting and snaring woodcock.

Winter Captures

The most successful method of capturing woodcock in the wintering grounds where large numbers concentrate in open fields is the use of a light and hand net at night. This method, used to catch birds for banding, requires great skill and is the most arduous.

Merovka (1939) was the first biologist to use a headlight and hand net to capture woodcock at night for banding. He and his associates in Louisiana banded 50 in January, 1937, and 15 in January, 1939. Norris (1941–42) banded 76 woodcock in 1940–41 and 28 in 1941–42. McIlhenny (1940a) used this method in 1939–40 to band 154 woodcock and 207 Wilson's Snipe (*Capella gallinago*).

Glasgow (1958) banded more woodcock between 1949 and 1958 than all the previous workers combined. With the aid of students and personnel of the Louisiana State Game Department he captured 8,192. During these years Glasgow refined winter capture techniques to a state of near perfection. He found a snowshoe-shaped net with a handle 12 feet long most efficient (Figure 3). The best light for shining the eyes of woodcock in the feeding fields was a hunter's headlight with a three-inch lens and operated from a six-volt hand lantern battery. An experienced crew of six has captured more than 250 birds in one night, and two experienced men occasionally captured 150 birds in one night.

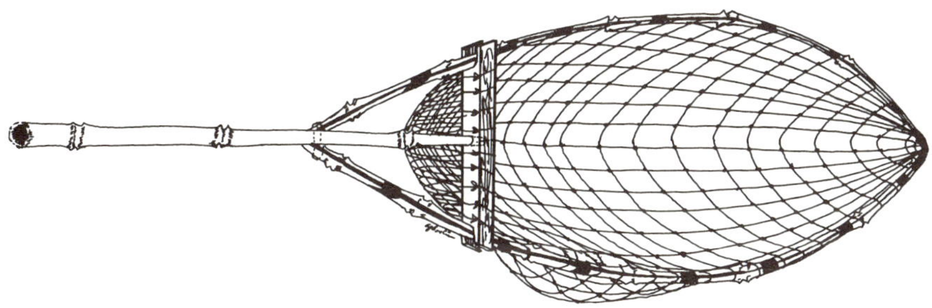

Figure 3. Glasgow's handnet

Glasgow found mild evenings in the dark of the moon during drizzly weather to be the most productive. In moonlight on bright nights the birds flushed wildly and relatively few were caught. In Maine, Martin (1964) and others successfully handnetted birds only on dark, wet nights.

In the South, birds feed in drainage ditches or other specific places in pastures or cornfields. Successful banders familiarize themselves with the configuration of a field in daylight before night netting.

STUDY AREAS

The unique character and habits of the woodcock have made it a focal point of scientific interest for many years, but woodcock are migratory while most research biologists are not. Early studies were based upon local observations by independent workers who were in touch with the bird only a few weeks of each year and rarely in contact with one another. Research on the breeding grounds was thus uncoordinated with that in the southern wintering range. True coordination began as recently as 1959 when the U. S. Bureau of Sport Fisheries and Wildlife assumed direction of the program and assigned biologists to conduct a range-wide study. Since I had been working on the breeding grounds since 1949, I selected a typical northern habitat, the Quabbin Reservation in central Massachusetts, with supplementary investigations in the Connecticut River Valley near Amherst. I made some additional observations in Vermont and Maine.

The Quabbin Reservation

Quabbin Reservation is 88,000 acres of land surrounding a reservoir of 35 square miles. This is a metropolitan watershed area which, except for boat and some shore fishing, is closed to the public. During the nineteenth century all of the woodlands in the region were cut down for railroad ties and lumber for charcoal, and most of the cleared land was devoted to diversified farming: rye, corn, and oats were grown, with grazing of upland areas by cattle and sheep. At the time the land was abandoned, farming was being conducted on a reduced scale; only 30 per cent of the area had been kept clear for crops and pasture. After Quabbin Reservation was established the remaining landowners were forced to sell their lands to the state and move elsewhere. The rising waters of the impoundment had claimed several small towns by 1941.

The Massachusetts Division of Forests and Parks, in cooperation with the Metropolitan Water Commission, began extensive reforestation for erosion control and watershed protection. Most of the open land was planted to red and white pine. When my woodcock studies began in 1950 most of the fields contained young evergreens, with many small openings comprising singing fields (Figure 4). The plantings had grown to young pine forests by 1960; most of the open areas formerly used by woodcock as singing fields had been filled by pine (Figure 5).

Quabbin, at present, contains a mixture of cover types, including

Figure 4. Woodcock singing field, Quabbin Reservation, 1950

Figure 5. Same field 1960 abandoned by woodcock

extensive stands of second-growth hardwoods, mixed forests of white pine and hardwoods, conifer plantations of different ages, and scattered swamps dominated by high-bush blueberry and alder. The reservoir in the Quabbin Reservation lies at a mean altitude of 524 feet above sea level, but elevations of the surrounding lands extend to 1,281 feet. The general aspect is one of wooded hills rising irregularly from the shores of the great man-made lake.

The reservation contains such varieties of soil types as Merrimac sandy loam, Gloucester loam, Gloucester stony loam, Gloucester fine sandy loam, and Whitman stony loam. If other habitat requirements are fulfilled, soil type in central Massachusetts does not influence woodcock distribution. Woodcock breed throughout the area wherever favorable openings exist.

Louisiana

The most important work on the wintering grounds has been conducted by Glasgow in Pointe Coupee, Iberville, and St. Landry parishes of Louisiana. These parishes are bounded on the west, north and east by natural levees of the Atchafalaya and Mississippi Rivers. Behind the primary levees of the main rivers a system of secondary levees follows the small streams which were formed by river crevasses.

The alluvial levee soils are fertile and have fair drainage. They become extremely hard when dry, especially on the back slopes where the finest clay particles were deposited. Much of the higher land is cultivated for sugar cane, corn and cotton, while the backs of these extensive levees, which have poorer drainage and adjoin the swamp, are often in pasture. All the study areas in Pointe Coupee and Iberville parishes were located on primary or secondary terraces. These fertile, alluvial terraces and adjacent extensive swamps provide woodcock with excellent winter habitat.

Iberville and Pointe Coupee parishes are typical of the overflow lands of South Louisiana and are very important woodcock wintering areas. Iberville Parish has a land area of 401,920 acres and and Pointe Coupee has 360,960 acres. While much of the land is forested, agriculture is important (Montgomery, 1956). The present trend is toward drainage, clearing and grazing. Grazing and the growing of cane and corn are the chief farming practices. All timber in both parishes is of bottomland hardwood-cypress types. Although oak, gum, and cypress are the predominant trees in the swamps, the woodlands contain many other trees as well as shrubs, vines, and briars.

Glasgow (1958) conducted some woodcock work in the upland parishes where pine was the dominant species of tree. As long as there is sufficient moisture on the hilltops to permit feeding, woodcock are distributed widely over the pine area. When moisture is deficient, the birds concentrate in creek bottoms, on seepy hillsides or at other similar moist sites. They frequently gather in young pine thickets, but are more abundant in creek bottoms. The hardwood-cypress bottoms found in the pinelands have a greater importance to woodcock than their land area would suggest.

CHAPTER II

History

The status of woodcock before the American Revolution is largely unknown. Most of the writers of that period ignored the species in favor of the larger, more conspicuous, and economically more important animals, such as waterfowl, wild turkey, and deer. The first known report of woodcock in North America was a passing reference by Paul Le Jeune (1634) who stated, as recorded by Pettingill, that "Woodcock were eaten by the Montagnois tribe of Iroquois in the vicinity of Quebec City."

Although there are few records to substantiate it, we may assume that colonial agriculture, with its patchwork of family farms and open-range livestock grazing, favored the woodcock. Locally and seasonally it was probably of more than passing interest to market shooters and sportsmen.

The anonymous author of *The Sportsman's Companion* (1783), the first American publication on game bird shooting, describes woodcock hunting methods in the late eighteenth century: "Good spaniels and boys to beat the cover with sticks, are the best to spring them; they are an easy shot, and generally arched, if you give them time to disentangle and clear themselves from the cover. . . ." The woodcock came into its own as a game bird a decade or two before the Civil War, and the shooting records of the sportsmen and market hunters of that time indicate an abundance of birds unknown to modern hunters.

Henry Herbert (1853), better known as Frank Forester, a competent observer of the American sporting scene, gives us a vivid account of woodcock hunting in 1839, remarkable because of the relative inefficiency of the muzzle-loading shotguns that were then used.

> On no ground, however, have I ever seen, or shall I—I much fear—ever again see this bird in such multitudes as on what are called the "Drowned Lands" in Orange County, New York. In 1839, I shot over it accompanied by my friend, Mr. Ward, who then weighed above three hundred pounds, and shot a single barreled Westley Richard's gun. In the successive days we bagged 57, 79, and 98 cock over a single brace of dogs, not beginning to shoot until it was late in the morning. In the following year, with a friend from New York, I shot on the same ground all day the first, and until noon on the second, bagging on the first one hundred and twenty-five birds, and on the second morning seventy. The first of these days was intensely hot; the ground became so much foiled by running of the innumerable birds that, although we had excellent retrievers, we lost beyond doubt, forty or fifty birds, and at 4:00 P.M. were entirely out of ammunition.
>
> I am perfectly satisfied that if we had been provided with a brace of fresh dogs at noon, with clean guns, and a proper supply of powder and copper caps, both of which gave out, it would have been perfectly easy on that day to have bagged from one hundred to one hundred and fifty *couple* of woodcock.

Herbert implies that woodcock had become scarcer by 1853, when his account was published. This assumed decline is verified by later observers. Lewis (1863) stated: "We all know it is no very unusual thing for even a tolerable shot to bag fifteen to twenty couples of young cocks in the course of one day's shooting." Bogardus (1874) wrote: "The most I ever killed in a day was fifteen couple. I have heard men boast of having killed fifty couple in a day but if they did it the birds must have been vastly more abundant than I ever saw them anywhere."

An early New England woodcock hunter, George Peabody (J. C. Phillips, letter), left a journal of his success while hunting in Essex County, Massachusetts. His woodcock bags suggest woodcock population trends. Between 1862 and 1872 he averaged 111 woodcock a year, with a low of 45 in 1865 and a high of 220 in 1867. Beginning in 1873 the birds apparently declined, and he shot an average of only 17 a year between 1873 and 1883. In 1880 his journal states, "I bagged 30 woodchucks shot through their heads with a rifle. I turned my attention in this direction because there were almost no woodcock." His success continued to decline until records indicate no woodcock killed in the early 1900's; the lack of success in the later years may have reflected less hunting effort, but no daily journals were found for these years—just his total bags. There are daily records through 1885.

By 1874, according to Pettingill, woodcock were selling in the New York City markets at $1.50 a pair; 1800 birds were sold by the markets in a single week. Market hunters on the wintering grounds in Louisiana usually hunted woodcock at night by "fire lighting" with the aid of blazing pine knots or oil soaked rags. They killed large numbers of birds in fields where they congregated at dusk for feeding. In later years they used more efficient carbide lights and battery-operated headlights.

There is some evidence that after the turn of the nineteenth century long open seasons and market hunting had taken a heavy toll of this famed bird. Fisher (1902) referred to it and the Wood Duck (*Aix sponsa*) as "two vanishing game birds." Sandys and Van Dyke (1904) wrote: "Good cock-shooting is now almost a thing of the past in many places where heavy bags were the rule. The heaviest bag ever I saw was sixty-odd to two guns in one day scored by two market hunters near Lake St. Clair. Extensive drainage, backed by too close shooting, has played havoc with leagues upon leagues of what formerly were the strongholds of this choice game." It was during this period, especially in the northeast, that solid stands of white pine invaded former woodcock habitat. The decline of deer populations paralleled diminishing woodcock populations. The woodcock and deer started their upturn after the cutting of pine got underway, to be replaced by mixed hardwood stands.

Scattered laws to protect woodcock were enacted by the more important woodcock states during the latter part of the nineteenth and the first few years of the twentieth centuries. However, the enforcement of these laws was so ineffectual that the woodcock gained little protection except that of the federal Lacey Act of 1900, which prohibited the interstate shipment of game killed in violation of state laws. With the enactment of more restrictive state laws and the improvement of state law enforcement, spring and summer shooting gradually waned. The Migratory Bird Treaty Act, which became effective in 1918, placed woodcock hunting under federal regulation and ended market hunting. By 1920 woodcock could be shot legally only in the fall. The bird responded well to this protection, and, while it never again attained the numbers that Henry Herbert knew, it has been abundant enough each year since 1920 to permit regulated hunting.

Information covering the past 25 years shows no significant trend, either up or down, in woodcock numbers. Singing ground counts, which have been made annually for the past 25 years, indicate that breeding woodcock today are only slightly less abundant than when the survey started. The only important deviation occurred in 1940 when birds declined as a result of a severe winter freeze in the South, but their populations recovered quickly. Neither has there been any noticeable decline in wintering birds over the past 15 years, but winter counting techniques are unreliable. The ebbs and flows of the annual wintering counts are largely attributable to variations in activity caused by changing weather conditions (Glasgow, 1958). Most reports on woodcock population trends are estimates and should be accepted with caution. Even in the case of counting males using singing grounds, a reported decline may reflect changed habitat rather than fewer birds.

Although there appear to have been no significant changes in continental woodcock numbers over the past 25 years, the abundance described in the nineteenth century may still occur when the birds concentrate locally on migration. I. Grant Scott (letter), hunting in Cape May, New Jersey, flushed between 75 and 100 woodcock on November 9, 1961. On November 11, he found 75 to 90 birds in half a day. A week later he reported that an even heavier flight than the first passed through Cape May, on the morning of November 18. George Weed and his brother flushed 68 woodcock in five hours of hunting in Minnesota on October 13, 1963. Edward Mikula, a Michigan biologist, flushed 86 birds in three hours in Clare County, Michigan, on October 29, 1959.

In 1962 I sent a questionnaire to all state and provincial game departments in the breeding range of the woodcock, asking whether their observations indicated an increase or decline in woodcock over the past twenty years. Seventeen of 39 inquiries were answered. Eight states reported populations unchanged, seven reported fewer birds, and three reported more. It is of interest, however, that most of the states reporting fewer birds suggested that deterioration of suitable habitat, rather than the gun, was responsible for the decrease. Liscinsky (1964) believes the gun was an important factor for decrease in certain sections of Pennsylvania. Many of the most important breeding areas, including New Brunswick, all New England states (except Rhode Island), and Wisconsin and Minnesota reported the populations unchanged. New York and Pennsylvania reported fewer birds. I have great confidence in the kill figures calculated by the Game Division of Michigan based on polling a selected sample of hunters. The kill increased from 49,320 in 1954 to 144,940 in 1964. At the same time spring census of breeding males in Michigan increased and woodcock flushed per hour rose from .35 in 1954 to .69 in 1964. These figures are based on thousands of hours of hunting. Ammann (letter) believes that this trend represents a "moderate" increase in birds and is not due entirely to an increase in hunting. In summary, my best estimate is that the woodcock is as abundant today as it was 25 years ago, although there are some threats to the population on the horizon which I shall discuss later.

CHAPTER III

Range

ANCIENT RANGE

In 1954 W. Auffenberg unearthed the first fossil bone fragment of an American Woodcock ever discovered, from Eichelberger cave near Belleview, Marian County, Florida (Brodkorb, 1957). Dr. Pierce Brodkorb (letter), comparing it to the stage of evolution of quail bone fragments in the same deposit, judged that the age of the bone dated to the middle Pleistocene. Pleistocene woodcock remains have been found at two additional sites since 1957. Ligon (1963) dug out 13 bone fragments of woodcock in Haile, Alachua County, Florida, and Wetmore (1962) found others at Natural Chimneys, Virginia. I know of no other ancient records of woodcock. These findings suggest that woodcock lived in America before the last two ice invasions, possibly more than 1,000,000 years ago. The modern bird probably evolved in the late Tertiary period as did many other present-day species of birds (Lanyon, 1963). The range of woodcock between the advance and retreat of four ice invasions of the Pleistocene period is unknown.

PRESENT RANGE

The woodcock range in North America has probably not changed substantially for hundreds of years. Our earliest records, which date back scarcely 200 years, contain almost no reports of the occurrence of woodcock where they do not exist today. There are isolated records of rare occurrences of woodcock west of their present range. In fact, the first record of woodcock in New Mexico was of a bird collected by Fred Turner on the Del Rio canal on January 25, 1964 (Harris, 1965).

While the woodcock has a wide geographical range in North America, this highly specialized bird occupies a restricted ecological niche. Most woodcock are confined to young forests with some scattered openings on poorly drained land. They are, for example, considered an abundant breeding bird in Nova Scotia, but I spent 52 months during 16 summers from 1919 to 1935 exploring and canoeing through the interior wilderness of that province, conscientiously keeping a list of birds observed, and I flushed only one woodcock. They are "a common to abundant breeding bird" (Mendall and Aldous) near the coast on the fringes of farming country from Yarmouth through southern Cape Breton and Cumberland Neck to New Brunswick.

Over 90 per cent of the woodcock on the continent are in an area bounded on the east by the Atlantic ocean, on the west by 95° W. longitude, on the north by 50° N. latitude, and on the south by 28° N. latitude (Figures 6 and 7). Breeding records of woodcock exist in every state and province encompassed by these geographical boundaries.

Most birds breed in northern areas of their range. A high percentage of the woodcock population breeds in eastern Minnesota, Ontario, Wisconsin, Michigan, New Brunswick, Nova Scotia, and Maine. Important states for woodcock breeding south of these are Ohio, New York, New Jersey, West Virginia, Pennsylvania, and all New England states except Rhode Island. Pockets of abundant breeding birds occur in the Appalachians as far south as North Carolina.

The precise range as described in the 1960 AOU Check-List is as follows:

> *Breeds* from southeastern Manitoba, western and central Ontario (Fort William, rarely; Tamagami), southern Quebec (to Gaspé) northern New Brunswick, Prince Edward Island, Nova Scotia (except northern Cape Breton Island), and southern Newfoundland (probably), west to central Minnesota, central Iowa, west-central Missouri, Arkansas (except extreme west) and extreme eastern Texas (rarely) and south to Louisiana, Mississippi, south-central Alabama, Georgia, and central Florida; formerly west to eastern North Dakota, eastern South Dakota, eastern Nebraska, eastern Kansas, eastern Oklahoma, and central eastern Texas (Houston and Hardin counties).

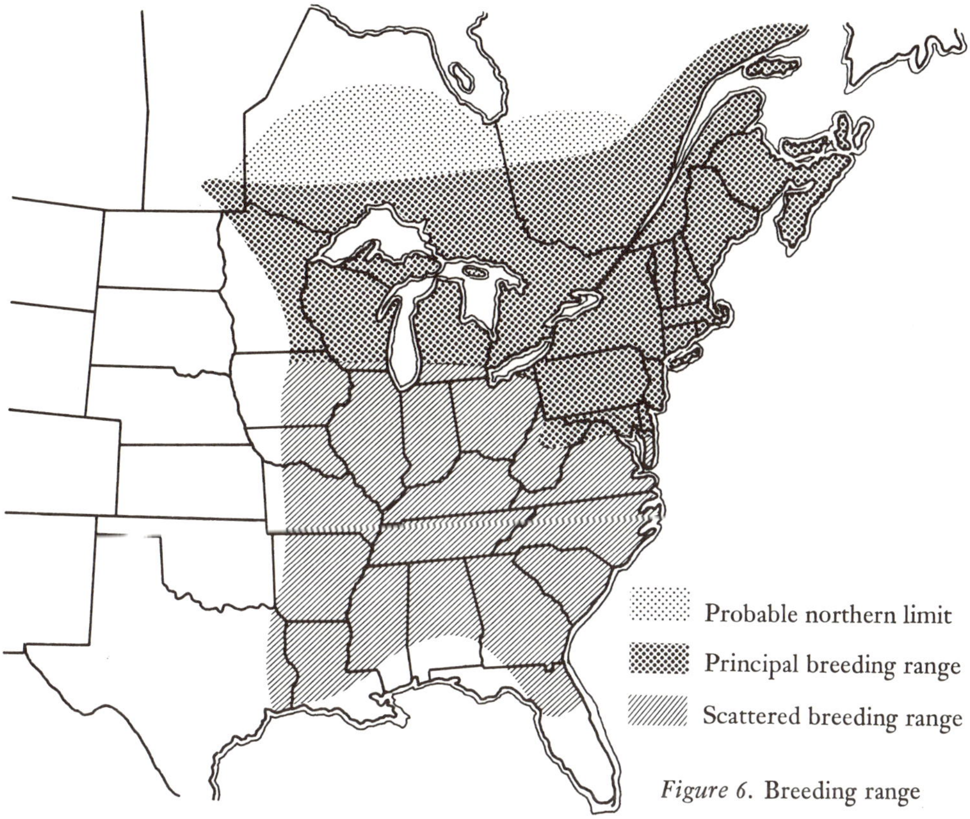

Figure 6. Breeding range

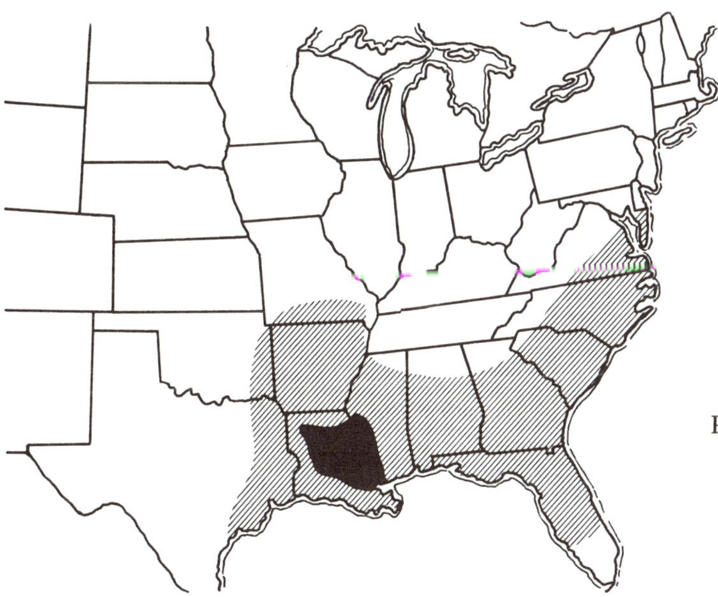

Population density

Figure 7. Winter range

> *Winters* mainly in southeastern Arkansas, Louisiana, and southwestern Mississippi; in lesser numbers from eastern Oklahoma, southern Missouri, southwestern Tennessee, northern Mississippi, northern Alabama, western South Carolina, west-central North Carolina, and southeastern and western Virginia south to eastern Texas (Montgomery, Harris, and Matagorda counties), southern Louisiana, southern Mississippi, southern Alabama, and southern Florida; rarely north to central Missouri and southern Illinois (near the Mississippi River), southern Indiana, southern Ohio, and western West Virginia (near Ohio River) and to New Jersey; occasionally in more northern parts of breeding range; accidental in Colorado (Tammath), Montana (Billings), northeastern Ontario (Onakawana), eastern Quebec (Betchewan) and Bermuda.

I believe by the more recent findings described below that the range should be extended north to northern Ontario and James Bay.

Northern Limits

In August, 1892, two canoes passed through heavy rapids on the Black River near Athabasca Lake in northern Saskatchewan. A quarter of a mile below the rapids J. B. Tyrell of the Canadian Geological Survey reported: "canoes were pushed in among the willows over a soft muddy swampy flat to the beginning of Woodcock Portage, so-called because we aroused a woodcock in one of the swamps as we passed it, this bird being exceedingly rare in the far north" (Preble, 1908). Preble (1902) quotes a report by Dr. Bell who saw a woodcock at York Factory at the end of August and says the woodcock may regularly reach southern Keewatin, but its presence as far north as York Factory must be merely accidental.

Clarke (letter, May 28, 1964) reports the occurrence of woodcock populations in the James Bay area and farther north. In part he says, "We have a spring record from the junction of the Ashweig and Winsk rivers just south of Hudson Bay. Farther east the birds are found quite regularly on the way to the southern end of James Bay. There are a few spring records but most of the observations at James Bay have been made during the goose hunting season when woodcock are commonly encountered in the willows along the coast." Cree Indians with whom I spoke at Ft. Albany, James Bay, said that they often see woodcock in the early fall. Wright (1956) records the occurrence of woodcock three miles north of Baie Johan Beetz on the St. Lawrence River and on Anticosti Island in the Gulf of St. Lawrence.

It is not surprising that nesting records are lacking for the northern extremities of the woodcock range, since finding nests or young birds is a matter of chance; Clarke reports large numbers of woodcock near Lake Abitibi, Ontario, in the fall, even though there are no breeding records from this area. In the vast wilderness of the North it is possible that woodcock breed in many suitable areas. The northernmost nest reported from Ontario was in Armstrong Township in June, 1958, north of Lake Superior and a few miles from Lake Nipigon (G. F. Boyer, 1959).

I wonder whether these northern records of woodcock represent an extension of range. Clarke has suggested "the possibility that there is a post-breeding northward migration." Although I cannot positively disprove this theory, I doubt it. Mendall and Aldous had two recoveries of woodcock banded as chicks and shot the next fall only six and seven miles, respectively, north of the point of banding. Of over 1500 woodcock that I banded during the spring and summer I have one fall recovery suggesting vagrant migration. An adult male trapped on a singing ground in late April, 1950, was shot the following fall in Westmoreland, New Hampshire, forty miles north.

In view of Preble's records, woodcock have probably ranged as far north as James Bay for years, but, with much improved methods of transportation, many more people who know woodcock have visited these areas and thus most observations are relatively recent. I doubt, therefore, that the recent reports of woodcock in the North can be accepted as proof of range extension.

Some investigators (Glasgow 1958, Liscinsky 1964) have found some correlation between woodcock distribution and the occurrence of earthworms. Although no systematic studies have been made of earthworm abundance in the North, the literature gives some conflicting clues (see p. 80).

Southern Limits of Breeding Range

Glasgow considers the woodcock to be a "regular though uncommon breeder" in Louisiana. Van Pelt (1951), Reid and Goodrum

(1953) have found nests in Louisiana. Howell (1924) classed the woodcock as a permanent resident in Alabama, but not abundant. Two more recent breeding records of the Fish and Wildlife service (Mendall and Aldous) are from Opelika and Decatur, Alabama. Pettingill recorded several instances of woodcock breeding in Arkansas, but, as in other southern states, it is considered rare in the summer. There are at least two breeding records from Florida, one near Tampa and the other at Jacksonville (Pettingill).

Winter Range

After the southern migration in the fall, woodcock tend to concentrate on a restricted winter range. The majority winter on the southern Atlantic coast as far south as Tampa, Florida (where two woodcock that I banded in Massachusetts were killed) and along the Gulf Coast to eastern Texas.

In Louisiana, Oberholser (1938) mentions that woodcock are "unbelievably numerous" in East Feliciana, West Feliciana, East Baton Rouge, West Baton Rouge, Livingston, Iberville and Pointe Coupee parishes. Mendall and Aldous add DeSoto, Sabine, Vernon, Beauregard, Jefferson Davis, Avoyelles, Ascension, Assumption, St. Tammany, Grant, Rapides, and Evangeline to the list of parishes that support a large winter population. Norris (1942) believed Pointe Coupee, West Baton Rouge, and Iberville parishes contained the most important wintering grounds in the state.

Range in Massachusetts

The ecological range of the woodcock varies considerably. I am intimately acquainted with range in Massachusetts, and since the state presents four entirely different ecological zones, I believe these details will be found fairly representative. The four zones are the spruce fir northern forests of Berkshire County, the transition forests of central and northeastern Massachusetts, the pitch pine scrub oak forests of Cape Cod, and the open heath and blueberry swamps of Nantucket Island.

Pettingill and also Forbush (1925) considered woodcock common in Massachusetts except in the three southeastern counties of Nantucket, Dukes, and Barnstable. It is possible that frequent fires which swept Dukes and Barnstable counties thirty years ago eliminated most of the woodcock habitat for many years. Dukes County now contains some of the best woodcock habitat in the state. Today the bird is a common breeder in Dukes and western Barnstable counties. I heard singing males in April, 1960, on Naushon Island, one of the Elizabeth chain. We banded several singing males on Martha's Vineyard and the chicks of two broods. This island continues to offer excellent early and late fall woodcock shooting even after most woodcock from the North have migrated through. On May 12, 1965, I made an experimental census run on the west side of Nantucket Island and counted 23 courting males in 3.5 miles. An average of over two birds per stop is one of the highest counts ever recorded in the woodcock breeding range. I do not know how many woodcock may winter on this 76-square-mile island. Those that migrate there would have to cross a minimum of 30 miles of open ocean by flying north to Cape Cod and south to Nantucket. If they flew directly north from Long Island, the flight would cross 100 miles of sea.

There is some favorable habitat in each county of the state, and I have found breeding woodcock in every county except Suffolk (the city of Boston). I counted 12 singing males in 1956 on three miles of dirt road bordering the Neponset River in Canton, seven miles from the center of Boston; the census route was bisected by several turnpikes.

Woodcock are most numerous in the western part of the state including Worcester, Hampshire, Franklin, Hampden and Berkshire counties. During my studies we found 250 singing grounds within 20 miles of Amherst in Hampshire County, and we probably did not cover more than 25 per cent of the potential breeding habitat.

With the help of technicians in the forestry department at the University of Massachusetts, I headed a team which made an ecological map of the state from aerial photographs; ecological niches as small as 10 acres were depicted (Sheldon, 1959). The composition, height, and density of all forests were mapped. We also classified abandoned fields growing back to forest, cut-over areas, abandoned and cultivated orchards, urban areas, croplands, shrub swamps, and five classes of wetlands. Approximately 625,000 acres of the

5,000,000 acres in Massachusetts were in early stages of forest succession, shrub swamps, or abandoned fields; I regard these areas as potential woodcock habitat. In 1850, 75 per cent of Massachusetts was in open croplands or urban settlement, but today about 75 per cent of the land is forested. With increasing cut-over lands, I estimate that woodcock range and habitat, although shifting locally, have not changed in ten years. Deterioration is certain to take place over the next twenty years as more and more rural areas are chewed by hungry bulldozers, and man's dwellings crowd the landscape.

HOME RANGE

Knowledge of woodcock home range is extremely meager except for the limited information obtained during the breeding season. Twenty-five per cent of breeding males in Massachusetts confine their courtship activities to one site. The land encompassing the singing field and daytime covert comprises the home range of most male woodcock in the spring. The floating males without territories probably range over wider areas than resident males. I found the distances separating diurnal resting and feeding areas from 55 singing sites of territorial males to average slightly over 100 yards. The distances varied from 50 feet to 440 yards. The male usually flew directly from his diurnal covert to his singing field and did not use the intervening land.

The incubating hen confines her range to the nest site and a nearby feeding area. Mendall and Aldous found the range of broods to be close to 100 acres until the chicks reach the flying stage. I found one Massachusetts brood in the same daytime resting place (May 1960) for ten consecutive days.

During the summer and winter between migrations, woodcock use a home range of several square miles. They fly at dusk from widely scattered daytime retreats to open fields where, given suitable weather, they feed until dawn before returning to diurnal coverts.

CHAPTER IV

Physical Characteristics

I can vividly remember an evening ritual of my family when I was a young boy. In October, when my father would come home from a day of grouse and woodcock hunting on the eastern fringes of the Green Mountains in Vermont, he would bring his birds home in a rucksack. He made a game of having three excited children open his bag to see what he had shot. Filled with curiosity, we would vie with each other to be the first to open the rucksack. I well remember my wonder when I first found a strange long-billed bird, for the woodcock is unlike any other bird. The woodcock is mottled brown, about the size of a man's fist, with a striped head the size of a golf ball; the shoebutton eyes are set high and far back on the head. Its most distinguishing feature is a bill two and a half inches long. It has short legs feathered to the heel. In its usual stance the head appears attached to the body with no neck. Closer examination of the color reveals a light cinnamon-colored breast and back and sides which appear to be washed with black and different shades and tints of brown, reminding one of a forest floor covered with dead leaves. Once seen, a woodcock is never forgotten.

PLUMAGE COLORATION

Males and females are colored alike: forehead and crown ashy gray; occiput with an indistinct median black stripe; nape clear black, with three or four narrow crossbars of deep buff or rufous; entire back, scapulars, and rump black, mottled and barred with rufous, but many feathers broadly edged or tipped with clear bluish gray; side of head ashy to buffy white, with a black line from base of bill to eye; a similar black stripe across the ear-coverts; chin white; sides of neck brownish ash; throat, breast and belly buff or pale cinnamon, deepening on the sides and flanks; primaries slate colored; secondaries and most of the wing coverts barred with black and buff; tail feathers mainly clear black, with ashy tips and silvery white below.

EXTERNAL SEPARATION OF SEXES

The male is three-quarters the size of the female, but sexes cannot be consistently distinguished when birds are flushed. Pettingill discovered that female woodcock are larger and heavier than males. Tufts (1940) was the first worker to suggest that bill length was a reliable sex criterion. Mendall and Aldous (1943) later determined from 188 specimens that bill length was the best technique for recognizing the sexes of woodcock. Bills of 72 mm. or over were always from females and birds with bills under 64 mm. were always from males. Birds with bills between 70 mm. and 72 mm. were females 95 per cent of the time, and birds with bills between 64 mm. and 66 mm. were males 95 per cent of the time. In birds with bills measuring between 67 mm. and 69 mm., sex was evenly divided. The amount of overlap (approximately 17 per cent) was much less than when sexing was based on weights or other measurements (Mendall and Aldous).

Greeley (1953) further refined sex separation by measuring the width of the outermost primary wing feather (Figure 8) two centimeters from the tip. He successfully sexed 97.2 per cent of 106 specimens, by gross examination of the outer three primaries. Harris (1953) used beak length and the combined width and length of the three outer primary feathers to sex woodcock. Blankenship (1957) followed Harris's method; by using the combined width of the outer three primaries and the bill length he reduced overlap to 0.36 per cent. Combined widths of the three outer primaries were 12.4 mm. or less for males and 12.6 mm. or more for females.

SCIENTIFIC CLASSIFICATION

The American Woodcock (*Philohela minor*) belongs to the Order Charadriiformes which includes the gulls, jacanas, oyster-catchers, plovers, turnstones and other surf birds, avocets, stilts, curlews, sandpipers, snipe, and phalaropes. Within this large group, the woodcock

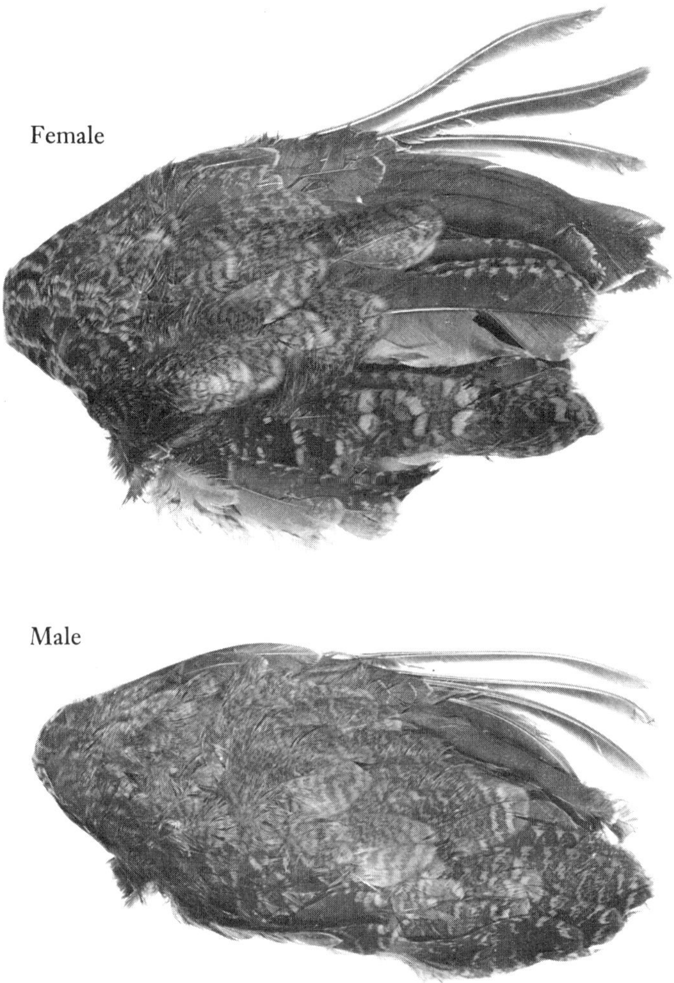

Figure 8. Sex differentiation by width of primaries (Photo by R. A. McCabe)

is classified in the family Scolopacidae containing such forms as curlews, willets, yellow-legs, godwits, snipe and sandpipers. The family is further divided into three subfamilies of which Scolopacinae includes the woodcock and snipe. This is the most specialized subfamily and is characterized by features of the skull and bill to be discussed below. Studies by European workers on the anatomy of the European Woodcock (*Scolopax rusticola*) will also be referred to since, with the exception of size, skull characters are identical with those of the American Woodcock (Mouseley, 1935).

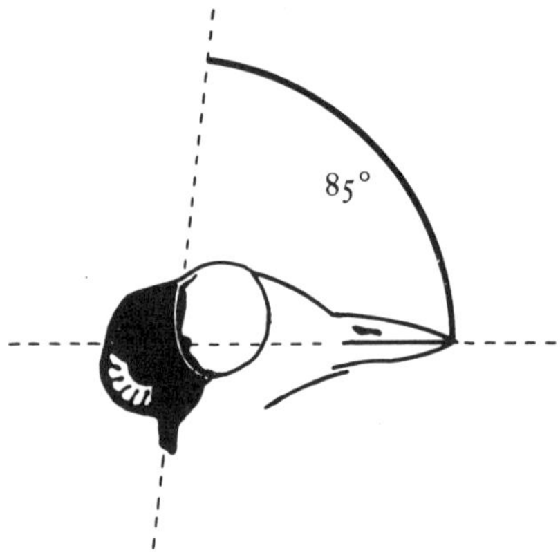

Figure 9. Cerebral axis

Woodcock, like ducks, have a long list of aliases. Among the most colorful are timberdoodle, bogsucker (parts of Canada), night partridge (Virginia), night peck (North Carolina), mudsnipe, blindsnipe, wall-snipe, whistler, brush snipe, siphon snipe, owl snipe (Nova Scotia), bog borer, hokumpoke, becasse (Cajun, Louisiana).

SKULL

As early as the latter part of the eighteenth century certain peculiar and extraordinary features of the woodcock skull were recognized (Mouseley, 1935). The best recent work on the American genus has been by Cobb (1959). The woodcock is the only bird with an "upsidedown" brain. The cerebellum of the brain is in a ventral position and connects with the spinal chord through a ventrally placed opening (the *foramen magnum*) with a slightly frontal slant. Cobb, comparing the brain positions of woodcock, Short-billed Dowitcher (*Limnodromus griseus*), Herring Gull (*Larus argentatus*), Semipalmated Sandpiper (*Ereunetes pusillus*) and Semipalmated Plover (*Charadrius semipalmatus*), measured what he termed the "cerebral axis." This is an axis plotted along the base of the brain, passing anteriorly through the middle of the olfactory lobe and posteriorly through a point in the middle of the brain stem below the caudal tip of the cerebellum (Figure 9). This indicated the amount of rotation of the brain backwards from its more usual anterior position in birds that have smaller orbits and shorter bills. The relative position of the woodcock brain is portrayed in Figure 10.

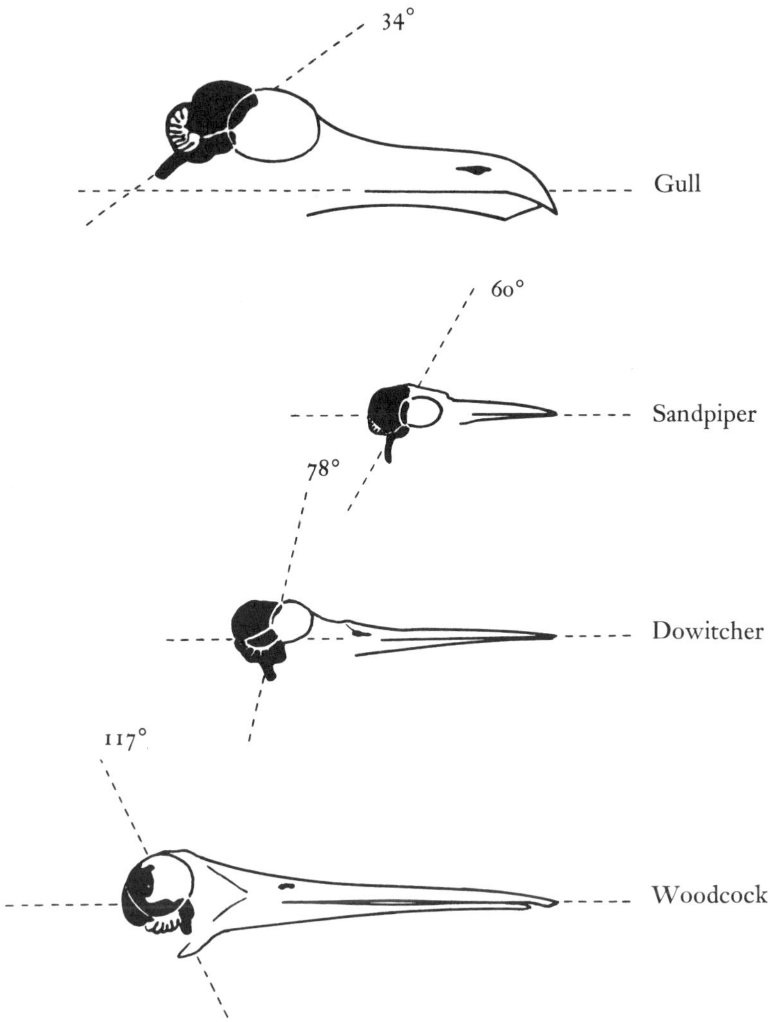

Figure 10. Comparative brain positions (Cobb)

The evolutionary development of the brain position suggests a highly specialized adaptation for food intake. Cobb, in speculating on the evolution of the woodcock skull, states: "The eye apparently moved backward as the bill lengthened and the nostril approached the base of the bill. As the eye developed its large size and posterior position, it gained space at the expense of the forebrain, which was rotated backward, causing the mid-brain and hind-brain to push downward and slightly forward. . . ."

The woodcock bill is from 2½ to 2¾ inches long, tapering in

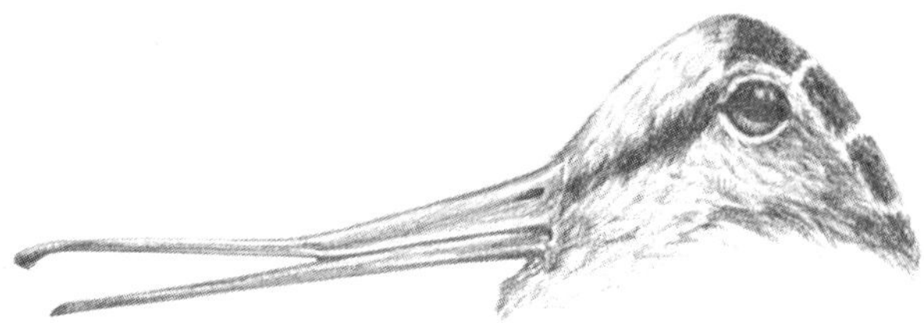

Figure 11. Distal end of upper bill

lateral profile. A membranous bill is the distinguishing characteristic of the subfamily Scolopacinae. The woodcock is equipped with a special bone and muscle combination which enables it to open the distal end of the upper bill (Figure 11). When the bird probes in the mud it can thus open its bill and grasp its prey.

SENSES

The senses connected with the bill and tongue—touch, taste, and smell—warrant some discussion since the method by which woodcock find food underground has long been a controversial question.

In 1775 Bowles described feeding behavior of captive European Woodcock kept in the aviary of Prince De Luis in Spain. This observer stated: "I did not observe a single instance in which it failed to get its prey, and both for this reason and because I had noticed that he never dug his beak further down than the orifice of the nostrils, I inferred that it was the sense of smell that guided him in . . . procuring his food."

Rowan (1931) has pointed out that in mammals the distance between nostrils and olfactory lobes contains a large tract of sensory epithelium responsible for picking up olfactory stimuli, whereas in birds the nostrils are not fleshy and sensitive. The cavity behind the nostrils of birds has few turbinal bones. Rowan concludes that with exception of the New Zealand Kiwi the nasal organ of birds is decadent. Allen (1925) states: "The only general conclusion that at present seems warranted is that many birds probably do have a certain power [of smell] but that they are little dependent on it."

Neuhaus (1962) has written the most authoritative modern book on the olfactory sense of birds. He thoroughly reviews studies made by many authors all over the world whose experiments appeared to indicate an olfactory sense in many species. He used some of the same methods as other workers and was unable in 1,242 trials to obtain evidence of learning in pigeon (*Columba livia*), Robin Redbreast (*Erithacus rubecula*), Blue Titmouse (*Parus caeruleus*), Thrush (*Monticola* sp.), and Grey Lag Goose (*Anser anser*). He concluded that earlier authors have not sufficiently disproved the possibility of visual perception.

I believe that the nostrils of the woodcock are close to the skull to enable the bird to probe with the full length of its bill. The end of the latter is covered with a membrane bearing the termini of many sensory nerves. Rowan refers to such a bill as "the most peculiar taste apparatus found in the tips of shore birds that probe in the mud. It is probably a combination of taste and touch." Allen (1925) calls the woodcock bill "a delicate organ of touch."

Dr. B. C. Wentworth of the University of Massachusetts made a microscopic examination of a cross section of the beak. It contains a concentration of blood vessels near the end, each of which is accompanied by a nerve ending. My own observations convince me that the nerves on the end of the bill are the woodcock's most important adaptation to detect prey under ground. I have watched feeding behavior of 162 captive birds held at different times over several years. Woodcock will often probe and hold the bill underground for a few seconds, then immediately extract it without a worm, walk rapidly a few inches and probe again, finding a worm. The tentative, searching probing may be repeated several times before locating prey. It is my conclusion that the nerve ends in the tip of the woodcock's bill act as antennae, detecting the movement of earthworms in the soil.

Other conspicuous characters of the beak are long protuberances, superficially resembling teeth, on the underside of the upper bill. The long tongue with its rugose surface, the rugosity of the upper bill, and the prehensile bill tip appear to be efficient adaptations for grasping and extracting slippery worms.

There is no bird in the avian kingdom other than the woodcock with such relatively large eyes placed so high and so far to the rear of the skull (Figure 12). As Allen (1925) suggests, it seems as if the bird must easily be able to see behind as well as above. He further suggests the large eye is an adaptation for dusk or night feeding.

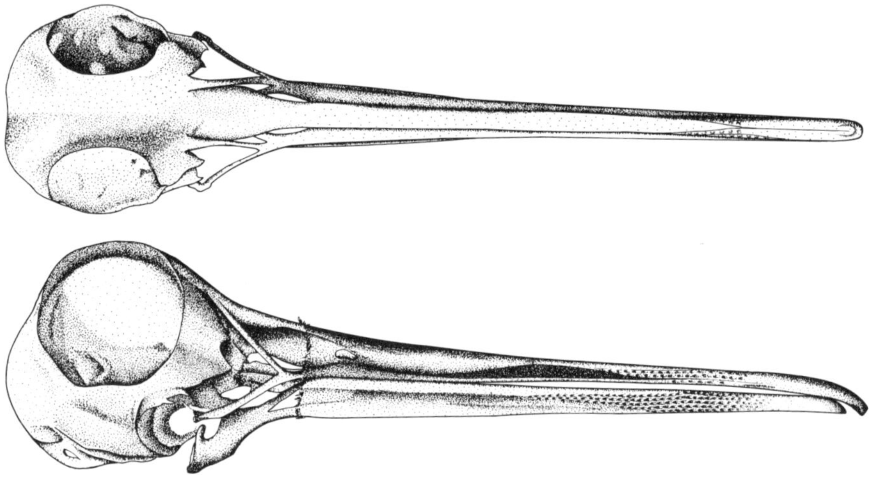

Figure 12. Skull of woodcock

Cobb writes that the size and the posterior position of the woodcock eye enable it to see more above and behind when feeding and that the position of the eye makes it "less likely to get in the mud."

There are many studies on the avian eye. Birds' eyes differ in the number of fovae (ultra-sensitive pits on the retina) and in their structure. No critical study has been made of the woodcock eye. Mouseley noted that the optic nerve and pecten of the eye of a jay correspond exactly with a drawing of the woodcock eye by an early English ornithologist (Cumberland) at the beginning of the nineteenth century.

When food is on the surface of the ground there is no doubt that the visual ability of the woodcock comes into play. During July of 1953 I watched a woodcock at dusk busily running around on top of a granite boulder feeding rapidly. After flushing this bird, I found by the light of a flashlight many very small ants on the rock. I have also watched woodcock chase flying moths that were attracted by lights in cages. Glasgow (1958) reports that on wet nights on the wintering grounds many of the worms consumed by woodcock are on the surface of the soil.

The woodcock ear is unique in its position on the head. It is below and slightly ahead of the eye; the ears of most birds are posterior to the eye.

Allen's (1925) description of the mechanism of the avian ear is pertinent. He says that "the hearing of birds next to sight, is the most important sense to them, and appears to be very keen. In the human ear the eardrums which vibrate to sound waves are slightly concave and connect with a series of three small bones named hammer, anvil and stirrup (from shape). In birds the eardrum is convex like a plug in the tube of the outer ear, while instead of a chain of three bones, there is a single bone, the columella, which acts like the dasher of a churn to transmit vibrations to the fluid of the inner ear; by this direct form of transmission birds are able to hear sounds we cannot." In World War I parrots were used to detect the approach of aeroplanes before they could be heard by the human ear.

I believe that the woodcock's hearing is secondary in acuteness only to its sight. Woodcock appear to differentiate, for example, the noise of a cottontail rabbit feeding nearby on the bird's singing ground, and the stealthy approach of a human intruder who keeps behind cover, invisible to the woodcock. In the former case it shows no alarm, while in the latter case it often flushes. The woodcock obviously recognizes calls or wing twitterings during the breeding season. Females respond to the notes of a male and fly into the breeding field. Males will flush and pursue a woodcock that is uttering breeding notes on a singing field it is trying to preempt. Whether or not hearing plays a part in feeding is controversial; I have been unable to find any evidence of this.

WINGS AND MOLT

The woodcock and the European Lapwing (*Vanellus vanellus*) are the only shorebirds with rounded rather than pointed wing tips.

I discovered, while netting hundreds of birds of both age groups during the summer, that juveniles molted all feathers except primary and outer secondary wing feathers in their first summer (Table A, Appendix II). Duvall (1955) identified juveniles before the summer molt by a color pattern on the tips of the feathers of the back and scapulars. The adults molted all feathers during the summer. Some adults started molting primaries as early as June 20, but took several weeks to complete the molt. The last feathers molted by the adults were the outer three primaries on each wing, which were molted simultaneously from the middle of August to early September.

We examined the tips of primaries 8, 9, and 10 with a binocular microscope and observed a characteristic wear which made it possible to distinguish the two age groups. The presence or absence of bursae of Fabricius (Greeley, 1953) enabled us to correctly age 95 per cent of 146 wings of known age, of woodcock from Wisconsin and New England (Sheldon, Greeley and Kupa, 1958). This discovery instituted the continental wing collections. Martin (1964) discovered that ages could be distinguished by macroscopic examination of the color pattern of the secondaries (Appendix IV), thus making it possible to age wings rapidly. Sex and age ratios of thousands of wings have been determined, thus providing the most complete data available for measuring annual productivity.

PROTECTIVE COLORATION

The colors of woodcock plumage are wonderfully adapted to blend with the bird's environment. In the evolution of the woodcock the development of the camouflage of feather colors was undoubtedly important to its survival. The woodcock escapes detection by its enemies by crouching on the forest floor. I have seen birds "freeze" in an open field at dusk when an owl swooped over; they act similarly in front of a pointing dog. Freshly-hatched chicks will remain completely still in whatever position they happen to be after the hen flushes. On May 6, 1959, I stepped on a chick before I discovered the brood in the open forest floor a few feet away. An incubating hen depends on camouflage and is difficult to flush from a nest.

VOICE

The woodcock has a wide array of vocal sounds, many of which have been described in the literature. Females are more vocal than males when first captured. Woodcock emit the following sounds:

1. The flight song. This has been given phonetically as a series of five or six *chips* during the descent (Pettingill). This phonetic rendition is misleading since the note has a liquid quality and might better be rendered as *chirp*.

2. A nasal, almost insect-like note commonly rendered as *peent.* This is uttered on the ground between aerial flights by the males. According to P. P. Kellogg (personal communication), "the *peent* is a complicated series of pulses. The repetition rate of the pulses (175/sec) gives the 'buzz' sound." The female often *peents* following copulation.
3. A two-syllable gurgling note best described as *tuko.* The male utters this note preceding each *peent* note. Whereas the *peent* can be heard several hundred yards, the *tuko* cannot be heard more than 50 feet from the bird.
4. *Cac-cac-cac-cac:* The very hoarse scolding note usually given in flight when one male chases another male from its singing territory. The notes are so close together that the sound is more like a buzzing. Two birds are always involved, and it is not certain that only males give this call.
5. *Chur-chur:* A low pitched nasal chatter heard from a female calling chicks which have been scattered.
6. Cat-like call: It is difficult to describe this sound, since it is completely unlike any other bird sounds I have heard. It sounds like a squealing cat, and has only been heard a few times in summer on the alighting fields. It is a loud, startling sound. Large birds, apparently females, have been flushed from the site of these sounds; there has been one direct observation by Dr. William Nutting of the University of Massachusetts who, concealed in bushes eight feet away, watched a female make this sound on a summer alighting field. This specific vocal call has not been reported in the literature.
7. *Quack-quack-quack:* To all intents and purposes this is similar to a duck sound with notes at much longer intervals than the *cac-cac-cac* notes; females caught in nets uttered this sound as they flew into the alighting field. Possibly males also give it, but I have no certain evidence of this. A variation of this heard on the wintering grounds has been described by Glasgow (personal communication) as similar to a lamb blatting, and is given by both sexes.
8. Squeal of hen with chicks: This sound, usually accompanied by the broken wing act, is very similar to the sharp cry given by a Ruffed Grouse (*Bonasa umbellus*) with young and may

be rendered as *wheeee.* I have made two observations of this in Massachusetts when a hen was flushed from a nest as the chicks were hatching. It is entirely unlike the cat-like squeal.

9. Flicker-like call: Captive birds use a subdued call which can be described as *wick, wick, wick.* I suspect that this indicates slight alarm.
10. Hawk-like cry: Birds in captivity often utter a relatively low drawn-out cry which can be rendered as *keeee.* I have heard this when I have been concealed, so probably it is not an alarm note.
11. *Wheeeep* is a distinctive high-pitched whistle-like note of distress by chicks up to 5 or 6 weeks of age.
12. Wing twitter. For many years ornithologists were not in agreement whether the accelerating, audible twitter heard in the ascent for the flight song was vocal, or made by the wings. In July, 1959, I removed the narrow three outer primaries of a singing male. He continued his flight song but the wing twitter was absent, indicating that the sound is non-vocal, made by air rushing through these feathers.

WEIGHTS

Many sportsmen contend that they can recognize "flight" (migrating) woodcock by their greater weight. Although this theory does not stand up well when exposed to unbiased scientific analysis, it is easy to understand how this impression might have arisen. The following information on weights will throw more light on this controversial subject.

Since the inception of my studies hundreds of woodcock collected throughout their range have been weighed at all seasons of the year. Early spring is the only season represented by a small sample. In early spring in Maine, Westfall (1954) weighed 46 males that averaged 127 grams (4.4 ounces). Eighteen spring males that Blankenship weighed in Michigan averaged 129 grams. Few females were weighed at this time of year, but it is reasonable to assume that both sexes reach their minimum weight after the long flight from the wintering grounds to the breeding grounds. Three females weighed by Blankenship in the spring averaged 167 grams. Fat reserves were

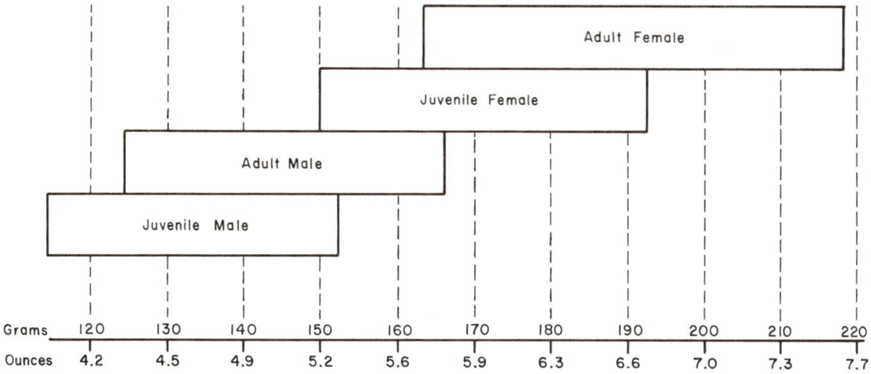

Figure 13. Summer weights of 152 woodcock captured in central Massachusetts, 1956–1959

used up, and the keel of the sternum was prominent in these specimens. One hundred and fifty-two woodcock that I captured in central Massachusetts showed weight gains from spring to summer (Figure 13 and Table B, Appendix II). Young birds which weighed from 13 to 15 grams when hatched grew rapidly and most had virtually attained adult weight within eight weeks (Figure 13).

During July and early August most adults or juveniles gain little, as attested to by the weights of the same birds caught at intervals during the summer (Table C, Appendix II). Specimens I collected in mid-August showed the earliest accumulation of fat. One adult

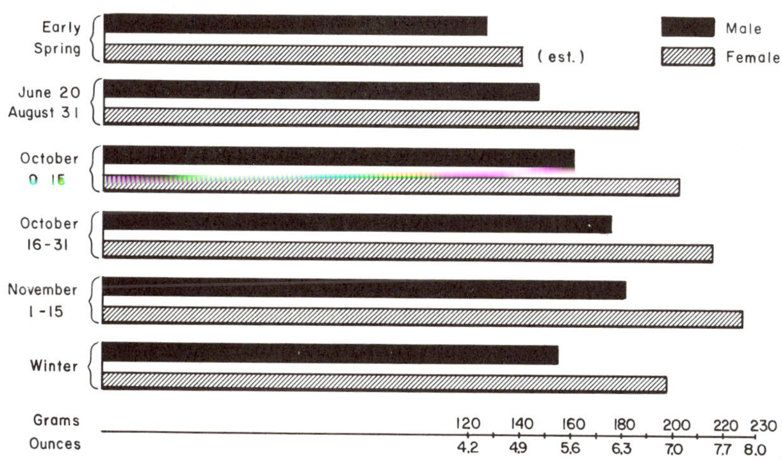

Figure 14. Mean seasonal weights of approximately 940 adult woodcock based on records of Maine, Massachusetts, Louisiana, and Nova Scotia

female, for example, captured on July 3, weighed 165 grams; when I recaptured her on August 28, she weighed 184 grams. Accumulating fat rapidly increases the weight through the fall. Robie Tufts (1940) classified weights of 179 Nova Scotia birds by two-week periods. In the first weeks of October males averaged 160 grams (5.6 ounces) and females 202 grams (7.1 ounces). During the first two weeks of November the males had attained 180 grams (6.3 ounces) and the females 226 grams (7.9 ounces) (Figure 14 and Table D, Appendix II). Blankenship found similar results when he weighed 180 fall-shot woodcock from Michigan and found that during the first two weeks in October males averaged 158 grams and females 196 grams. During the first nine days in November males had gained 25 grams and females 35 grams.

Occasionally there is some sex segregation in migration. Since the late-fall birds weigh the most, the chances are greater that late-flight birds would in fact be heavier, especially if there was a preponderance of females. "Native" birds would be equally heavy if they were still present.

Table E, Appendix II, presents fall weights of woodcock from two Canadian provinces and four states. The birds from New Brunswick are lighter, because the hunting season there opens on September 15 before birds have gained much weight. Most of the other birds were killed in October. The Nova Scotia birds were taken in late October and early November and the Vermont birds within the first 20 days of October. The heaviest bird we have recorded is a 275 gram (9¾ ounces) female shot in Nova Scotia (Tufts, 1940), indicating that females occasionally exceed one-half pound.

The only fall weights available segregated by age and sex are those provided by Marshall (personal communication) from Minnesota and Greeley (1953) from Wisconsin (Tables F and G, Appendix II). The combined data from these investigations reveal there is no significant difference between weights of juveniles and adults in the fall.

Winter weights as recorded by Glasgow (1958) and by Reid and Goodrum (1952) (Figure 14), average 6 to 7 per cent lighter than maximum fall weights and are significantly heavier than spring and summer weights (Table B, Appendix II). In periods of extreme cold in Louisiana, when many woodcock are forced to the coastal parishes, severe winter weight losses and considerable starvation may occur. In the freeze of 1951, for example, Lynch (1951) weighed 12 females whose average weight was 167.5 grams. Assuming that the usual average winter weight of female woodcock is 199 grams, those birds lost on an average 15.8 per cent of their normal weight. In 13 males, Lynch found a weight loss of 17 per cent, assuming 156 grams is the average weight in normal winters. Other freezes have caused similar effects; that of 1940 is discussed by McIlhenny (1940), Lynch (1951), and Mendall and Aldous. Similar severe storms have been described for the south Atlantic coast by Wayne (1899), Smith (1901), Rice (1924), and Gregory (1893).

It is remarkable that so little weight is lost in the fall migration. My interpretation is that the woodcock's southern migration is leisurely and woodcock do not burn up the fat reserve to the same extent as some shore birds which sustain long non-stop migrations. Most passerine birds also appear to lose fat reserves on the southern migration (Wolfson, 1945).

CHAPTER V

Breeding Habits

COURTSHIP

Accompany me on an early April evening to an abandoned orchard near Red Brook in central Massachusetts. As twilight approaches, the western sky fades from a rich pink to a light gold. A ghost-like bird, reminding us of a giant moth, flits by, silhouetted against the sky, and alights with a flutter in an opening 40 feet away. We are audience to one of the most remarkable performances in the avian world—the courtship sky dance of the male American Woodcock.

Immediately after the bird alights we hear the nasal, insect-like peent note of our visitor. Through binoculars we watch him walk to and fro and, at 20-second intervals, tilt his head to the side and, with an apparently convulsive effort, open his beak to utter his characteristic call. Each peent is preceded by the soft gurgle-like tuko, uttered with bill closed. After 158 peents, occupying five minutes, our bird flushes, quickly rising on a gradual plane until we see him against the twilight sky. Fifty feet above his departure point on the ground the accelerating wing beats create a musical twitter. At first in wide circles that encompass three acres, he rises higher and higher in ever-smaller circles or ellipses until he appears bat like 275 feet above us. Briefly the twittering of wings stops, and he appears virtually to hover as he pours forth his true song—a variable series of liquid chirps, which he repeats several times as he glides to the ground on a zigzag course. He lands close to the same spot from which he had launched himself a minute before. After repeating his ground call for two minutes he again launches himself skyward. For 35 minutes we witness this performance; he makes 14 flights. Once a visiting male from a neighboring singing ground interrupts him, but he chases the intruder off, uttering a harsh and threatening cac-cac-cac!

Figure 15. Early woodcock (Etching by A. Lassell Ripley)

Just before his last flight in the gathering darkness, we hear a flutter of wings as another bird alights after its low silent approach. The male ceases peenting, but still uttering the tuko notes, raises his wings and walks with stiff legs to the female. We can hear his wings flutter as mating takes place on the ground. Almost immediately he ascends for his last flight and the hen flies silently from the field.

Brewster (1896), Pettingill, Mendall and Aldous and countless others have described the courtship antics of the male woodcock. The bird, usually an inconspicuous recluse, advertises his presence on his singing grounds in the spring; thus his evening displays have attracted the attention of ornithologists for many years. In spite of the wealth of literature on this phase of woodcock life history, I believed that enough remained unknown to warrant detailed studies.

Counts of singing woodcock have been the census method for estimating annual population changes. The results have played a part in determining the length of hunting seasons and the daily bag limit—further justification, I felt, for intensive studies.

Polygamy

Pettingill and Norris, Beule and Studholme (1940) believed the American woodcock to be polygamous. Mendall and Aldous concluded the bird was monogamous. Our trapping results in Massachusetts certainly indicate the male bird is at least promiscuous. No observations were made of a pair constantly together or cooperating in rearing young, as takes place with another member of the woodcock family, the Wilson's Snipe; the latter is monogamous, with the female occasionally participating in the flight song, and each caring for the young (Tuck, 1965 letter).

In Massachusetts I retrapped 48 birds a week or more after the initial capture, of which 23 were recaptured in different singing sites during the same spring; these capture sites for birds caught as repeats were from ¼ to 3½ miles apart. At six singing fields two males were caught on the same site the same evening. One hundred and twenty birds were captured the same spring in 60 singing fields. I trapped five different males on five successive nights early in April 1950 at one field, but, in this case, migration may have been responsible. After the spring migration in 1952 one of my graduate students trapped five males in the same decoy trap in a three-week period. Several males were captured at three and four widely separated sites the same spring. One bird was caught 2½ miles away from his usual site one night and returned to his "home" territory the following night. Most birds, captured more than once, tried to copulate with decoys.

I have often been asked if disturbance by trapping may have prompted these movements. I believe that this cannot be ruled out in all cases, but 16 repeats not included in the 48 above were birds caught on the same trapping sites on consecutive nights. Twenty-five of the 48 birds recaptured at intervals of a week or more had not changed singing grounds. Most singing sites from which males had shifted the same spring had been pre-empted by a new bird.

On two occasions I changed the location of a decoy in the same field on successive nights and observed the same male attempting copulation on both nights.

Many males paid no attention to a decoy in a courting field even when alighting within three feet. There may be obscure physiological factors governing mating. I believe it is reasonable to assume that a male will mate with more than one female which may visit his field and, furthermore, that he will not hesitate to move to other fields in the same spring for breeding.

Male Replacement

Stewart and Aldrich (1951) made a study of breeding male replacement of passerine birds on a 40-acre tract of a spruce fir forest in northern Maine, finding 148 territories occupied. They systematically collected as many males as possible for three weeks. For most species, more than twice as many adult males were collected on the area as were present before the collecting started. When males were shot, "the resultant voids in the population were filled by new males almost as soon as the voids were produced." The indications were "that a large, surplus, floating population of unmated males must have been present. . . ." Lack (1946) found that "when the owner of a territory disappears, the owners of the neighboring territories expand in the vacant site almost at once. . . ." Nice (1941) states that "birds which fail to obtain territory, form a reserve supply

Figure 16. Male approaching decoy (Photo by Richard Borden)

from which replacements come in case of death of owners of territory."

The replacement of male birds in passerine monogamous species is amply documented, and there is some evidence that this phenomenon also occurs in polygamous species. Bump, *et al.* (1947) discovered that Ruffed Grouse "tend quite rapidly to move into a covert whose population has been depleted out of proportion to those surrounding it." Hochbaum (1944) speaks of the surplus of unmated drakes in several species of ducks: "The unmated males . . . are active roving birds. They show no attachment to any given area, but are 'on the go' most of the time. They are opportunists, ready to pursue and court any unmated female. . . ."

Lumsden (personal communication) shot 12 singing male woodcock on Manitoulin Island, Lake Huron, after the spring migration; the sites from which the males had been removed were reoccupied by other males within a day. I netted and removed two males the same night at four singing sites between May 5 and May 13, 1964. These singing grounds were all reoccupied a week later when singing in late May had begun to decline. Two fields were of several acres and the other two less than one acre. It is not known if woodcock grounds are reoccupied by yearling birds or by males which had failed to find satisfactory singing territories. There appears to be a drifting opportunistic male population which promptly takes over abandoned singing sites. My studies, revealing that males will change singing territory, suggest the further possibility that there are preferred singing grounds; the replacement males may have been driven off originally, and occupied less attractive sites.

Copulation

Male woodcock often try to copulate with decoys. My field assistant, Ethan Howard, and two graduate students observed males mating with live females. The behavior with decoys and live females was the same. A few male woodcock saw the decoy while alighting in the singing field before their first flight. They did not peent but gave the tuko note at intervals. Raising their wings (Figure 16), they approached the hen at a direct, uninterrupted stiff walk or made a rapid series of dashes. After reaching her, each fluttered to her back with its wings beating slowly at first and then rapidly as the mating act was completed (Figure 17). Other birds spotting a

Figure 17. Male copulating with decoy (Photo by L. A. Campbell)

decoy seemed highly stimulated and often made five or six flights in rapid succession before copulation. When alighting they sometimes approached the head of the decoy and appeared to dip their heads as if bowing. Some tried repeatedly to copulate with some decoys set in the field without a trap. The bird shown in Figure 17 copulating with a decoy ignored a live female which had fluttered in close by. Other members of the sandpiper family, such as the Upland Plover (*Bartramia longicauda*), raise their wings in a similar way while courting.

One might construe the approach with the wings elevated as a "strut." Pettingill cites several ornithologists who reported seeing the male woodcock strut like a "turkey cock" with his tail erected and spread, but we did not record this in our studies. Norris, Beule and Studholme (1940) observed male woodcock approaching mounted decoys. Their account was similar to ours. They did not see a bird strutting, although on two occasions they saw males spread their tails in a fan shape over their backs. Pitelka (1943) observed a male bird elevating and spreading his tail on a singing ground. Figure 18 shows a male woodcock a moment before copulation. The wings were dropped from an elevated position to a point so low

Figure 18. Male before copulation (Photo by L. A. Campbell)

Figure 19. Tail fanned in alarm (Photo by R. A. McCabe)

they almost touched the ground, an action almost like a turkey's. I have watched woodcock erecting and spreading their tail feathers under the following conditions: A female flushed from a nest with newly hatched young; a male on a singing ground when he detected my presence; wounded woodcock; captive woodcock when I entered their pen; males trapped in a decoy trap and on the ground in front of a dog just before the bird flushed. I believe that in most cases this behavior is one of alarm (Figure 19).

Vocal Notes in Singing Fields

Male woodcock have a repertoire of four vocal notes during their courtship performance. On the ground they utter the peent and tuko notes. Brewster (1894) observed that the male opens his bill with each peent note. I concur in this, having often seen it in favorable light. One graduate student in 1964, using a pair of binoculars in an open singing site, counted a bird opening his mouth 58 times when peenting. The tuko note precedes each peent; it is often repeated several times without a peent, especially when a male approaches a female. The true song of the woodcock is a series of musical chirps emitted during his courtship flight. The so-called cackle note is a warning uttered by a male when another male invades his singing territory. I have trapped and netted several cackling males attracted by a tape recorder set up on their singing ground and playing the peent call. One of these males attacked a decoy. F. W. Davis (personal communication) saw one male chasing another while cackling; the birds collided in the air, both fluttering to the ground, but neither was injured. This note is more frequent in early spring when males are competing for singing sites, but it is given in areas of high breeding intensity throughout the spring. Although my studies suggested that this scolding call is used only by males, a recent observation by Fred Payne in Maine (1962) indicates that females may also cackle. Payne was able to imitate the peent call. In most cases singing males would fly over him giving the cackle note. In one case, however, a cackling bird flew from surrounding alders; he shot it and identified it as a female. In summer I have netted females which were uttering a quacking note in the air before capture.

Although Mendall and Aldous made no positive observations of females peenting, Pettingill stated that the female sometimes peents. Westfall (1954) reported two cases (based on circumstantial evidence) of females peenting. Norris, Beule and Studholme (1943) also reported evidence of females occasionally peenting. Massachusetts studies indicated that females sometimes peent, although no peenting bird believed to be a female was collected. A large number of captive females were very vocal, on occasion giving a quack-like note very similar to a peent. It is all but impossible to distinguish females from males in daylight, let alone in semi-darkness. Therefore sexes can be separated only on the basis of behavior. Martin (letter) collected 50 peenting birds in Michigan. All were males which suggests that females seldom peent. The circumstantial evidence on two occasions, however, was convincing. We observed copulation, the male went into a flight song, and the female remaining on the ground peented. The field notes of two students are given below:

> The male bird had completed eight flight performances when a female bird flew into the singing area. The moment that the female bird arrived on the singing site, the male immediately raised its wings and approached the female rather slowly at first, followed by

pursuit at a rapid pace. This went on, being confined to only about seven square feet, for about 30 seconds, when the male went through a flight performance. Immediately after the male left the singing ground for the first time following the female's arrival, the female began peenting. This note was not unlike that of the male. It lasted for only a short period of perhaps 30 seconds, and was highly irregular. After the male returned, he raised his wings, lowered them and went through a second flight performance; returning to the site he pursued the female and copulation took place. It was characterized by a slow wing flutter, followed by an extremely rapid wing beat, occupying a few seconds. After the mating was completed, the female flew off. The male continued to go through the familiar flight performances for some time after the female had left the site. The male made 20 flights.

April 1952 (R. Smith)

On May 13 binoculars were used for a clearer view of the activities of the singing male. The male arrived on the singing ground at 7:29 P.M. E.S.T. After peenting, he went into a flight song which was followed by three additional flights. While in his fifth flight, a female flew silently into the field. She gave a series of tuko notes. The male landed a few yards away, peented four times and went into a flight. He alighted ten feet from her, peented seven times before she walked toward him. Copulation took place, with the male fluttering his wings. Immediately afterward she flew about 20 yards away and landed still in sight. The male flew skyward for another song. Ten seconds after he flew she peented twice and continued making the tuko note. The male made three more flights but stayed on the ground very briefly peenting, seven, three and four times respectively. After the tenth flight, he peented three times before she flew within ten feet of him. They both walked toward each other, and a second copulation took place. She moved only five feet away and the eleventh flight took place. She was carefully watched with binoculars, and was heard to peent twice while the male was in flight. Her peents resembled the male's almost exactly. The male landed near her without peenting, but approached for a third copulation. . . .

May 13, 1952 (A. Pelletier)

Pelletier witnessed a fourth copulation and the female peented twice more.

This male was watched for 37 evenings between March 31 and

May 26. On the evening cited above, he made the greatest number of flights (18) recorded for any single bird all spring on nights without moonlight.

Height of Flight

I have not found in the literature any precise measure of the height of a woodcock in courtship flight, although estimates vary from 200 to 500 feet. In the spring of 1963 I measured three flights of an individual bird. I established a base line on the ground and stationed an observer with a modified plane table and vertical-angle measurer developed by Hagar (unpublished) for measuring the altitude of migrating hawks from each end of the line. We took sights of the height of three consecutive flights. All measured 275 feet. More measurements must be made to find the variation in altitude reached by different males.

Effect of Weather

Goudy (1960) studied factors influencing singing ground counts in Michigan; he found that singing activity declined significantly when temperatures fell below 41° F. Blankenship found the critical temperature below which singing decreased to be 35° F. I found that stormy weather, including high winds with or without heavy rains, downpours with no wind, and temperatures below freezing all curtailed breeding activity. Birds performed with equal vigor on cloudy or clear evenings or in light rain and mist. On April 15, 1950, on a clear night with a temperature of approximately 32° F, after a snowfall of April 14, birds sang actively, lighting on bare ground between snow patches and sometimes on the crusted snow itself. On May 8, 1963, I counted 17 birds on one of our census routes during a light rain. I found the same number on this route on May 9 when there were clear skies and no wind.

Although Goudy (1960) and Duke (1964) did not find that the moon affected flight performances, my Massachusetts studies demonstrated that bright moonlight could influence courtship activity. Birds sang later under full moons, and studies of four individual color-marked birds by graduate student Lewis Cowardin between

April 4 and May 14 in 1959 showed erratic behavior during the height of the moon; the birds were more active and changed singing grounds during the evening. During the full of the moon, one bird which had been using the same site all spring was caught two and one-half miles away, and the following evening he returned to his "home territory." Birds perform all through the night with bright moonlight. Blankenship also found that birds behave erratically on moonlight nights. The studies of Goudy (1960) and Duke (1964) suggest that moonlight will not affect census counting significantly.

Duration of Performance

Leopold and Eynon (1961), in 24 readings of light intensity, found that woodcock give their first evening peent between 5.0 to .02 foot candles. The mean, with standard error, was .71 ± .26 foot-candle. Light varies according to topography. Birds in deep valleys will start singing before woodcock on the tops of hills with western exposures. The length of performance, the number of flights, the interval spent on the ground, the number of peents vary during the same evening with individual birds. I found that, with the exception of the evening starting time and average duration of each flight, no two woodcock performed the same; no one bird duplicated in detail his singing behavior each night even under similar weather conditions.

On May 8, 1961, I stationed observers familiar with woodcock courtship at 13 singing sites in the Amherst area. Performance time from the first to the last peent ranged from 30 to 49 minutes. The courtship of most birds lasted a different length of time, and the number of flights varied from 9 to 18. Two of the birds performed for 37 minutes, and two for 39, but none of the remaining birds performed for the same length of time (Table H, Appendix II). Similar tests with the same results were run on April 16 and May 7, 1962, and May 13 and 14, 1963.

In 1952 I assigned one student to a study of a single singing male from late March to late May. This bird was no more consistent in his activities each evening than the birds cited above. With a stopwatch we timed 428 courtship flights of this woodcock. The longest flight was 68 seconds and the shortest 40 seconds. Fifty-six per cent of the flights lasted between 55 and 62 seconds (Table I, Figure 51, Appendix II).

The bird spent between 15 and 39 minutes on the ground (Figure 52, Appendix II) and from 6 to 21 minutes in the air during 35 evenings of similar weather conditions. His evening courtship activities varied from 48 minutes to 27 minutes before the mating season began to end in late May; the average number of peents uttered between flights each evening varied from 20 to 125 (Figures 53 and 54, Appendix II).

Dawn Flights

On May 13 and 14, 1963, we timed 8 woodcock in the evening and dawn under similar weather conditions; it was overcast with a 10 mile an hour wind. The average performance time at dawn was 59 minutes, in the evening 32 minutes; three dawn singing periods lasted 70 minutes (from first to last peent), and eight lasted more than 52 minutes (Table I, Appendix II). The census taker might assume from this that dawn would be a better time to count male woodcock than evening because of the longer performance. However, there were 41 minutes between the earliest and the latest hour when woodcock start singing at dawn; it would be difficult to standardize the time to start the count. I believe evening counts are the most reliable for census purposes.

Latitude

Woodcock performing in the most northern breeding areas sing the longest. Wright (personal communication) finds that most males in New Brunswick sing at least 45 minutes, but in central Maryland courtship performance lasts only about 20 minutes. Male woodcock begin to sing as soon as they arrive in Massachusetts at the end of March and continue until May 20, when singing gradually falls off. Courtship has ceased by June 1 in most years. Farther north, the breeding season begins about a week later, and in such southern states as North Carolina and Kentucky, breeding starts a month earlier and ends by May 1.

Peaks of Singing Activity

Westfall (1954), on a census route in Greenbush, Maine, in 1953 and 1954, counted the highest number of singing males in May (Table J, Appendix II). Similar peaks of activity in May were recorded by Norris (personal communication) in Newburyport, Massachusetts, and by Blankenship (1951) in Michigan. In 1951 and 1952, with the help of graduate students I ran an experimental census route stopping at .2 mile intervals. In three runs between April 15 and April 26, 1951, an average count of 10 singing birds was tabulated; three runs between May 1 and May 8 yielded an average of 19 birds. In 1952 the differences between comparable dates were 11 and 13.4 birds. If we combine the average of these two years, we see that the May peak is conspicuous (Figure 55, Appendix II). In 1953 the same route was run at intervals of .4 of a mile. In 11 runs between April 15 and April 30 a mean of 6.4 birds was counted, and in seven runs between May 1 and May 11 a mean of 9.8 birds was heard. One peak came before most hens had started incubating in April and the other after most broods were brought off in May.

In an effort to find the reason for the second peak, I stationed experienced observers at five singing grounds in Massachusetts in the spring of 1962. On April 16, with a temperature of 41–42° F, a three-quarter overcast sky and a slight wind, the observers simultaneously observed five singing males. There was a half-moon behind clouds with enough open sky to approximate the light intensity of a clear night with no moon. We counted all birds heard peenting in the vicinity, the number of flights of the individual bird being observed, and the total duration of the flight performance.

On May 7 we again investigated the same five singing grounds, recording the same data as in April. The temperature was identical, the sky clear with a one-quarter moon and a slight wind. We could not be certain that the same birds were performing on the same grounds as in April, but it is probably no coincidence that the number of flights and length of performance followed the same pattern. The average performance in April lasted 24.4 minutes (from the first flight to the last peent), and in May, 38 minutes. The average number of flights in April was seven and in May, eleven. In April we heard one additional bird peenting on only one of the five fields.

In May we detected six additional birds at some time during [illegible] evening on all the fields. If we had run a census of these five fields on April 16 we would have tabulated six singing males, and if in May, would have recorded a maximum of eleven.

We do not know the reason for this increased activity late in the singing season. The only correlation in late April and the first two weeks of May in Massachusetts may be with the height of the hatch (Figure 21). I suggest that late nests hatching in late May and as late as June 18 are renests; further study may reveal a correlation.

Flights in Other Seasons

Mendall and Aldous found two male woodcock performing "half-hearted" courting flights in late July. Pettingill cited instances of woodcock mating flights in October, but he did not hear the vocal song of the typical spring flight. Liscinsky (1964) heard a woodcock going through a typical flight song on October 18 in central Pennsylvania. I have witnessed many flights in July in Massachusetts similar to those described by Mendall and Aldous. Birds captured when descending from such flights proved to be juvenile males. On some summer feeding grounds males have often been observed approaching females with raised wings, similar in every way to the mating approach in the spring, except that copulation did not occur. Birds collected while engaged in this display had regressed gonads; their testes were not examined for evidence of active sperm. Woodcock have been heard peenting occasionally in every month from February to December in Massachusetts.

Two graduate students (T. Ripley and W. Fitzpatrick), while duck hunting on Cape Cod in late December, 1952, heard woodcock making typical courtship flights at dawn, and Latham (1956) watched the courtship flights of three male woodcock near Dover, Delaware, on January 8, when the temperature was 25° F. In late October, a graduate student saw a large group of woodcock leaving an alder swamp in Massachusetts just before daylight, many of which had been peenting while still on the ground.

The extent of courtship and breeding in Louisiana at the southern edge of the woodcock's range confirms other observations that, during the northward migration, males engage in courtship activities if the weather is favorable (Glasgow, 1958). Glasgow heard males making courtship flights as early as December 2; it is possible that some start singing prior to their arrival in Louisiana. He heard 12 woodcock singing in December, and banded several males and flushed other woodcock on December 7, 10 and 21 in fields where he had heard only one singing male. He banded 16 males in one field where two males were singing earlier the same evening. Thus it appears that early in the winter only a small percentage of the birds show evidence of the onset of courtship. The singing seemed to Glasgow to be "half-hearted"; it lasted only a short time and was rather sporadic. One male made two aerial flights and then peented on the ground for seven or eight minutes, another made four flights, uttering eight to sixteen peents between each ascent, and still another peented for about three or four minutes and did not fly. As a rule, the birds made one to three aerial flights, peented on the ground a few times, and then remained quiet.

In January the males sang longer and more often. Glasgow (1958) counted 43 males singing in January. Several of these birds made more than five aerial flights and peented eight to twenty times between flights. Toward the end of January the frequency of singing almost reached that heard on the northern breeding grounds, some individual birds performing for 30 minutes. Norris (1941) heard woodcock singing all night near Hammond, Louisiana, on February 12, 1941. Glasgow heard eight males singing in February in a field from which five were banded and fifteen other woodcock flushed. Whenever clear warm weather prevailed during February, courting males could be heard in the evening. He caught a singing male on three different nights within 75 feet of the same spot. In winter, singing woodcock seemed to prefer pastures containing stands of yankee weed three to five feet tall with openings 25 to 40 feet in diameter. They also used the borders of tall weed patches, fence rows, and openings in briars or shrubby growth.

Territoriality of Courting Male

Territoriality in animals is a controversial subject and most of the workers in this field are unanimous in agreeing that much more intensive investigation is necessary before we can interpret territorial behavior with confidence (Tinbergen, 1936 and 1957, Nice, 1941). Tinbergen (1957) describes bird territory as "the outcome of two distinct tendencies each of which may occur without the other, viz. site attachment and intraspecific hostility." He discusses the possible survival value of site attachment, for example, on the theory that familiarity with a site enables a bird to take advantage of escape cover quickly. Arguments have been put forward that show intraspecific hostility is one of many dispersing mechanisms employed in nature and that it is a highly efficient one. Nice (1941), in an exhaustive review of the literature on the theory of territorialism, classified territorial types as follows:

1. Mating, nesting and feeding ground for young.
2. Mating and nesting but not feeding ground.
3. Mating station only.
4. Restricted to narrow surroundings of nest as in such colonial species as Herring Gulls (*Larus argentatus*).

Much of the discussion of territorialism in birds concerns monogamous species. In polygamous species, such as woodcock, Ruffed Grouse, Ring-necked Pheasant (*Phasianus colchicus*), Wild Turkey (*Meleagris gallopago*), and others, only the male seems to exhibit territorial tendencies. Mayr (1935) states that "territory is an area occupied by one male of a species which it defends against intrusions of other males of the same species and in which it makes itself conspicuous." I do not believe that any of these definitions precisely describe territorialism of the male American Woodcock, but Nice's type 3 and Mayr's definition come the nearest to fitting the woodcock.

Brown (1964) has made a recent study of territoriality in birds to assess factors causing territoriality. He says the "aggressive behavior is generally employed by individuals in the acquisition of goals which tend to maximize individual survival and reproduction." He makes the further point that such behavior must be due to competition for a feature vital to the survival of a species. It must further be a factor which possesses what he refers to as "economic defendability"; to oversimplify, this means that a bird must be able to compete efficiently whatever the limiting factor may be. There may be competition for food, nesting sites, mates, roosting spots or breeding space. The most important of these, which through evolution insures the survival of the species and for which there is competition, is the one or more which a bird defends.

Let us examine the woodcock against this background. Wetherbee (personal communication) and I suggest the following observations on this bird. The woodcock is a sandpiper that has become adapted to the forest. Its unique physical adaptations and behavior can be explained as adjustments to a habitat remarkable for a shorebird, but it never lost the vestigial requirement for open space for courtship and breeding. I believe that the important factor in survival of the woodcock is competition for breeding sites. The following attributes of woodcock might be directly or indirectly correlated with competition for the short supply (at least historically) of courtship sites in the forest: territoriality restricted to the opportunity to copulate; polygamy, minimal sexual dimorphism; position of eyes; flight song in twilight.

The forest opening is defended by the male with peculiar calls and aggressive behavior. The cac-cac-cac note is used by a male in pursuit of an intruding male. Suitable openings, in northern woodcock habitat, are fewer than the number of potential breeding males present. Openings suitable in our eyes may not be accepted by woodcock because of subtle habitat features we have not recognized. Trapping or netting on singing sites has consistently yielded more than one male; these openings are defended by the male because they are the asset most necessary to reproduction.

The peent is made for communication on the ground in the dusk, particularly under circumstances where visual cues would be ineffective. The frequency of this note varies from 3 to 6 kilocycles but has components as low as .1 kilocycle. The low note has the greatest carrying capacity. In contrast, other members of the sandpiper family lack this low note. Examples include: Pectoral Sandpiper (*Erolia melonotes*) 2½–3 kc., Stilt Sandpiper (*Micropalama himantopus*) 1¼–2¼ kc., and Upland Plover 1½–3 kc. The flights with auditory cues (wing twitter and song) are a most effective advertisement by the claimant in general defense of the opportunity to breed. I know of no evidence of territorial behavior of woodcock hens and broods, or competition for feeding sites.

Since the male is polygamous, individual females would not be "economically defendable." Therefore I believe the defense of the open space has been the factor exploited, and evolution of the exhibitional and aggressive behavior has resulted. Unlike diurnal birds that maintain their territories by means of bright plumage or a watchful perch, the woodcock, pre-adapted for strong flight and non-perching, has exploited the sky to achieve conspicuousness. The male advertises himself to females by his silhouette in the twilight sky, and by sounds. His territory includes the area encompassed by his aerial flight.

The peculiar position of the woodcock's eyes, in view of the bird's feeding habit, may have evolved purely as a defense against approaching enemies. There is, however, a less obvious advantage for the female in being able to see the aerial antics of her suitor regardless of her position or current activity, but probably the necessity of each male's knowing the flight location of every other male is the salient evolutionary fact.

In the Arctic many shorebirds breed as early as possible, partly to insure that their offspring will develop strength in flight before bad weather precludes easy availability of food, thus providing the fat reserves necessary for the long southern migration. This cannot be true of the woodcock, but it seems significant that most breeding takes place before leaves of deciduous trees close the tree crowns. Dense foliage in coverts would make the aerial flight of a male invisible to a female. This may add weight to my belief that with woodcock the third dimension is important in their territorial behavior. In short, the courtship behavior and morphology of this shorebird of the woods have evolved to maximize the chances of reproduction for the species.

SINGING GROUNDS

Size and Shape

A singing ground is the site of the male's courtship performance; it may have several adjacent alighting spots. According to Mendall, Aldous, and Blankenship the range of sizes and shapes of singing grounds is endless. In Massachusetts I found singing sites as small as 100 square feet and as large as several thousand square feet. One universal requirement is a "get away" route for the bird's aerial

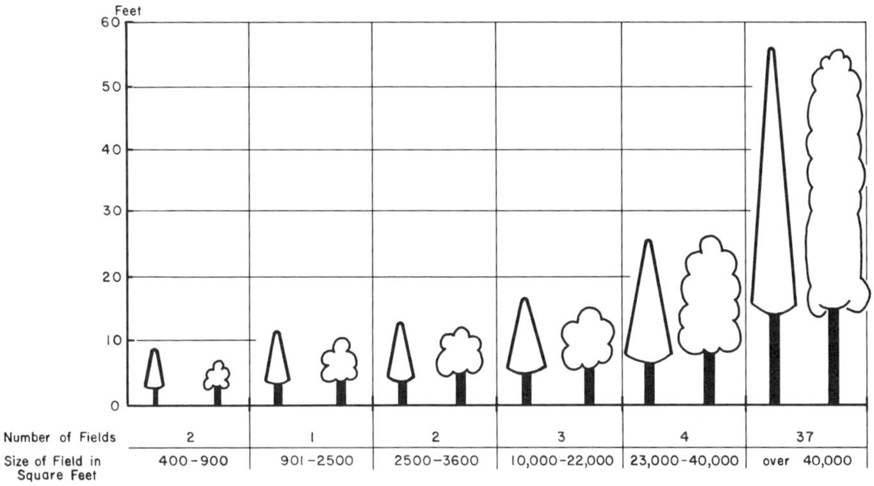

Figure 20. Tree height and size of singing field

flight. The male woodcock starts his flight with a fast flush, flying as low as five to ten feet for about 15 yards before gaining altitude. High surrounding trees, therefore, may limit the usefulness of an otherwise good singing site. In a study of 17 singing grounds in Minnesota, Dangler and Marshall (1950) found that three sites of 70 to 130 feet in diameter were encircled by trees up to 30 feet, and the remainder were sites of more than 130 feet with surrounding trees taller than 30 feet. Maxfield (1961), in a study of 50 Massachusetts singing grounds, found the height of surrounding trees tolerated by woodcock varied directly with the size of the courting site (Figure 20 and Table K, Appendix II). A few sites in Massachusetts were on slopes as steep as 45 per cent, but the actual alighting spots were usually on small, relatively flat benches.

Vegetation

Maxfield (1961) and I made a detailed study of the vegetation of 50 singing grounds in Massachusetts. I suspected that some clue to the preference shown by singing males for certain fields might be revealed by plant indicators. Certain plant species kept cropping up in the tabulation; bluestem grass in varying amounts occurred in each site, and meadowsweet was common in most of the fields. Gray birch, aspen, white pine, alder, red maple, dead grasses, the red cedar-ground juniper mixture, black cherry, or panicled dogwood, and other shrubs and young trees were the commonest shrubs on courtship sites. Gray birch and aspen usually invade bare soil or burned areas. These species germinate and thrive only on mineral soils or on very light sod (Hawley and Hawes, 1912). Alder was the characteristic plant on wet areas. In some abandoned pastures red cedar and ground juniper were the distinctive woody species; both can germinate and survive on a well-developed sod (Hawley and Hawes, 1912). Courtship takes place only when there are scattered woody plants one to two feet high, in early succession. Most herbaceous plants are prostrate during the woodcock mating period, but the alighting spot is often a finger of herbaceous cover extending into the shrubs.

Nantucket Island, Massachusetts, where I made the highest count of breeding birds in Massachusetts, averaging over two birds per stop, has unique cover. This island is no more than a flat, triangular spit of sand about 19 miles long and 4 miles wide. Vegetation consists of a stand of pitch pine at one end of the area, beach grass, heath, and small knolls covered with a brushy mixture of small pines, cedar and scrub oak. There are acres of open abandoned fields, and many small areas are swamp, grown up to high-bush blueberry. These appear to be the focal points of woodcock habitat. I have never seen so many woodcock so far removed from forest land. The birds were found several miles from the stand of pine. Elsewhere, I once watched a male woodcock fly from his daytime cover over cultivated, clean pastured fields, crossing a fence to reach a brushy singing site. Except for a few sites on dirt roads, two in coarse grass in large fields, and one on the edge of a cornfield where the stalks had not been cut, all sites that I found contained woody cover.

Dangler and Marshall (1950) and Marshall (1958) found a few scattered shrubs or some type of early stage woody plants on 47 singing grounds in central Minnesota. Marshall estimates that when woody vegetation grew six to ten feet high over more than 60 per cent of an area, woodcock would abandon it. Blankenship, in studying 98 singing grounds in Michigan, states: "The most important vegetational characteristic seemed to be that plant succession on the openings was in the early woody or low shrubby stage." The kinds of woody plants and herbs, except for those which may indicate soil type or moisture content, seem to be of little importance.

Marshall noted that most of the Maine singing grounds pictured by Mendall and Aldous were very similar to the Minnesota grounds. On the other hand, conifers form an important component of Massachusetts fields. In Pennsylvania, Norris, Beule and Studholme (1942) found birds performing on small openings in the "barren grounds" which are characterized by pitch pine, scrub oak and aspen cover. Plant species that have been recorded on singing grounds of different regions are shown in Table L, Appendix II.

Diurnal Cover

Mendall and Aldous found, in 24 instances, that the daytime cover of the male was located within 100 yards of his singing ground. We measured the distances from the singing ground to 55 diurnal territories of Massachusetts woodcock. Sixteen were immediately adjacent to the singing fields, where males walked to the courting site. Thirty-three were situated 150 to 200 yards from the singing site. Four were between 230 and 440 yards distant, and two more than 440 yards. In early spring males fed on a number of singing fields which were wet enough to support earthworms.

Soil Types

Marshall (1958) reported that 47 singing fields in the Cloquet Experimental Forest in Minnesota were located, for the most part, on loam and alluvial soils. Blankenship analyzed the soil in 80 singing sites in Michigan, finding that 29 sites were on sand, 10 on loamy soil, 16 on sandy soil, 17 on loam, 2 on muck and 6 on peat. In Massachusetts woodcock usually established singing grounds in abandoned fields with sandy loam soil. There were some exceptions: three birds of over 200 observed used abandoned gravel pits, and some used gravel roads.

Origin of Massachusetts Singing Grounds

The 50 Massachusetts singing grounds intensively studied were created directly or indirectly by man. Thirty-six were in abandoned fields, three in orchards, three in pastures, two on power line rightsof-way, four on dirt roads, one in a former burn, and one in cutover land. In the days before colonization, woodcock undoubtedly used natural openings in the forest—areas along stream beds, beaver "meadows," clearings created by fire, hurricanes, or Indians, forests destroyed by insects, and beach grass sites near the coast. I found birds using the latter areas on Martha's Vineyard, Massachusetts, within sound of ocean breakers.

Longevity of Singing Sites

The life of a singing opening depends on local land use and the prevailing vegetation. Fields of low-bush blueberry or ground juniper, for example, persist longer than openings in cutover forest, one blueberry field in Massachusetts having been used at least 15 years. If high vegetation is continually brushed or cut, singing grounds can be maintained indefinitely.

In wild areas, singing sites are usually dynamic and are useful to woodcock for only a limited time. Many of the first singing grounds that I observed were in abandoned fields planted to red and white pine in Quabbin Reservation. These plantings have now grown up to forests. Of 67 singing grounds that were being used in 1951, only 16 were occupied in 1960 (Figures 4, 5). Longley (1957), studying 16 singing grounds on the Whitewater Game Refuge in Minnesota from 1951 to 1957, found that when reed-canary grass invaded singing grounds they became unacceptable to woodcock.

NESTING

The most striking characteristic of the nesting woodcock is its reluctance to leave its eggs. A forester who was planting seedling pines in an open field on the Quabbin Reservation in May, 1954, told me that he had come within a single inch of impaling an unnoticed woodcock hen on his planting bar. The bird flushed only when the heavy iron tool struck the earth so close that it cracked one of her eggs. On May 27, 1953, I flushed an incubating hen one foot below the nose of my English setter. This characteristic behavior and the effective camouflage of the bird's plumage make woodcock nests notoriously difficult to find. The discovery of an incubating hen is

Figure 21. Early woodcock nest (Photo by Michigan Conservation Department)

largely a matter of chance even when crews of assistants are available for a systematic search. Well-trained bird dogs find a few nests under ideal scenting conditions.

Construction and Cover

A woodcock nest is very rudimentary compared to those of many other birds. The usual nest is a cuplike depression on the ground with a few twigs arranged around the rim. The hen makes no effort to conceal it; woodcock plumage is such effective camouflage that close inspection is required to recognize an unconcealed, incubating hen.

The bird often makes her nest at the base of a small tree or shrub (Figure 21). I have found woodcock nests in abandoned fields, conifer plantations, brushy areas, mixed forests of different ages, and blueberry fields (Figures 22, 23). Thirty Massachusetts nests were within 50 yards of an edge. Mendall and Aldous classified the habitat of 228 Maine nests. Forty-four per cent of these were in mixed

Figure 22. Prime nesting cover (Photo by Eldon Clark)

growths of birch, aspen, other hardwoods and conifers; 26 per cent in alder; 21 per cent in hardwoods, and the remaining 9 per cent in brushland, blueberry land or old fields.

Distance from Singing Grounds

Woodcock usually nest within 150 yards of an occupied singing field, although Weeden (1956) found one in Maine that was nesting ¾ of a mile from the singing ground where she had been trapped and color-marked in the spring. Mendall and Aldous found that 115 Maine woodcock nested an average of 345 feet from the nearest occupied singing ground. Blankenship, in Michigan, found the average distance for 31 nests to be 320 feet; in Massachusetts I found the average distance for 9 nests to be 310 feet.

Eggs

Woodcock eggs have been described in detail by Bent (1927), Pettingill, and other ornithologists. They are a pinkish buff to cinnamon, covered with small dark brown spots or splotches. They are remarkably large for such a small bird and are close to the size and weight of the eggs of Ruffed Grouse. The hatching weight of woodcock eggs (based on three eggs) averaged 15.5 grams, and Ruffed Grouse (thirteen eggs) averaged 14.5 grams at hatching. Eggs of both species have a volume of 15 cc. (Wetherbee, 1961). Measurements of 53 eggs in the United States National Museum averaged 38 by 29 millimeters, ranging from 35 x 27.5 millimeters to 41 x 40 millimeters (Pettingill).

Nest Success and Clutch Size

Nine of the 30 nests found or reported in Massachusetts are known to have hatched successfully. Five were destroyed by some predator, and seven were abandoned because they were disturbed by man or dog in early incubation. The fate of the remaining nine was not known. Each nest contained four eggs.

Mendall and Aldous gathered egg data on 115 nests in Maine. All except seven contained four eggs, the exceptions being late nests with three eggs. They found fertility was 98.4 per cent, hatching success 67 per cent. Most hens that lose their first clutches probably

Figure 23. Over-mature nesting cover (Photo by Eldon Clark)

renest. Mendall and Aldous felt it a conservative estimate that 75 per cent of woodcock hens produce successful clutches.

Blankenship stated that 31 Michigan nests contained two to six eggs with an average clutch size of 3.9 (Blankenship's record of six eggs is unusual). Lincoln (1951) reported a woodcock nest containing 12 eggs found by Paul Byron at Wheeler Reservoir, Decatur, Alabama, on April 30, 1958. I suspect this was a dump nest where more than one hen laid eggs; Byron did not discover whether any of these eggs hatched.

Hatching Dates

Average annual hatching dates for woodcock may vary as much as 14 days in the same latitude, depending on weather during migration and breeding. The hatch occurs at progressively later dates from the southern limits of the breeding range to the northern. A woodcock chick a few days old was found in Louisiana on January 29 (Beyer, *et al.* 1956), and Russell (1958) saw seven broods between April 6 and April 30 in Kentucky. Early records from other parts of range (Pettingill) are: Louisiana, January 29; Georgia, February 4; Virginia, February 24; New Jersey, March 9; Connecticut, March 24; Maine, April 17. Mendall and Aldous found that the average

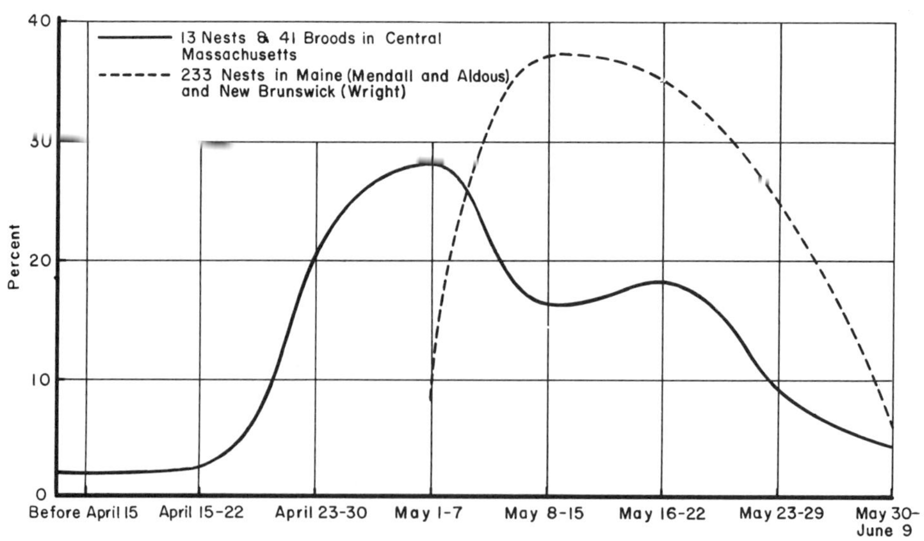

Figure 24. Hatching dates, 1950–1961, known and estimated

Figure 25. Incubating hen (Photo by Wilfred Kimber)

hatching dates of 173 Maine nests over six years was between May 11 and May 25. Hatching dates for New Brunswick near Fredericton were about the same as eastern Maine (Wright, personal communication). During my study, fifty-two per cent of Massachusetts nests hatched before May 7 and 69 per cent by May 15. The earliest Massachusetts hatching date which I have recorded is April 9 and the latest June 15. Hatching dates in Maine, New Brunswick, and Massachusetts are depicted graphically in Figure 24 (Table M, Appendix II).

Behavior of Incubating Hen

A hen woodcock lays one egg a day until the normal clutch of four is complete. If she is disturbed during the early stages of incubation, she will usually abandon the nest. In the late stages of incubation, she consistently returns to the nest after being flushed. Pettingill observed that incubating hens commonly leave their nests at dusk, presumably to feed, and return after dark.

On an afternoon of May 5, I flushed a hen from a nest in which the first egg had been laid April 11, 25 days earlier. Two chicks were hatched and already dry and the other two eggs were pipped. The hen flew 20 feet with her tail depressed and legs hanging. She alighted on a nearby road and feigned injury, crying not unlike a Ruffed Grouse hen disturbed with a brood. I left the scene and returned an hour later; the hen had returned to the nest, where I flushed her again. The two pipped eggs were hatched, but the chicks were still wet. I banded the four hatchlings and left, returning to inspect the nest 15 minutes later; by that time she was brooding the chicks. The following day she had taken the chicks off, and even with the aid of a dog, I was unable to find the brood. In May, 1963, I repeatedly flushed a hen from a nest containing eggs ready to hatch. Concealing myself, I watched her return to the nest; she walked across a dirt road in front of me. Her gait was measured, with the waltzing and bobbing motion usual when a woodcock is nervous or disturbed. When she reached cover across the road, she sneaked to the nest, reoccupying it 20 minutes after I had flushed her (Figure 25).

BROODS

Although the European Woodcock often has two broods a year, with two peaks of singing activity in April and July (Steinfatt, 1938), the American bird has only one. If the American Woodcock has two broods I would have recognized two age classes of juveniles in the hundreds of woodcock netted on summer fields. Woodcock estimated to be five weeks old were only netted in late June or early July; they still gave the wheep note of a chick. I identified juveniles throughout July and early August by plumage characteristics. From mid-August to mid-September all the juveniles were molting, and by late September most of them had completed their molt. If there were two broods, molting would be evident in late October or early November.

Chicks

The head and neck of an unhatched chick is folded under the right wing. At the time of hatching, the chick pips a hole at the obtuse end of the egg with its egg tooth. Anchoring the bill against the

Figure 26. Freshly hatched chicks (Photo by U.S. Fish and Wildlife Service)

rim of the hole, the chick by muscular contraction arches its thoracic and cervical vertebrae against the imprisoning shell until it splits longitudinally (Wetherbee, 1962). This lengthwise splitting is unique among birds.

WEIGHT AND GROWTH

Woodcock chicks weigh between 10 and 15 grams at hatching. Growth is very rapid. On a small island, Pettingill found a brood which he was able to contact daily. The young gained 5.3 grams during the first three days, 32.2 grams in seven days, and weighed 55 grams when two weeks old. At the end of four weeks the juveniles are almost full-grown, of strong flight, and indistinguishable from adults except for slight plumage characteristics.

BEHAVIOR

Woodcock chicks become very active soon after hatching. If the hen is disturbed she may lead them as far as 100 yards from the nest

Figure 27. Fleeing juvenile woodcock

(Mendall and Aldous) when less than two days old. In 1964 I collected a woodcock egg and hatched it in an incubator. The chick survived 80 hours after hatching on the food reserves of the egg yolk. I captured a newly hatched chick and held it for the first few days in a cage having an earth floor in which worms were buried; by the third day after hatching the chick was busily probing and finding worms in much the same manner as an adult bird. I believe woodcock hatchlings feed themselves as soon as the egg yolk is resorbed.

From the day of hatching young woodcock "freeze" when threatened with danger. Their mottled downy plumage is effective camouflage (Figure 26). I have found many broods of chicks which have developed primary wing feathers at 7 to 11 days of age. Although "freezing" is the common defense, chicks attempt to escape if discovered. Raising their wings vertically for balance (Figure 27), they run rapidly, scattering in all directions. I have noticed that they use the elevated wings to help them in hopping over logs or other impediments in their escape route. They can be caught quite easily if one can keep them in sight.

During the first few days after hatching the hen broods her chicks frequently, especially during inclement weather. After their primaries have grown brooding ceases. With increasing age the young birds, when feeding, wander progressively greater distances from the hen. I found one brood scattered over half an acre when they could fly short distances and were estimated to be three weeks old.

The brood stays together for varying periods; I have frequently

Figure 28. Hen with chicks (Photo by U.S. Fish and Wildlife Service)

seen four young birds capable of flight resting in a tightly packed group. My best estimate is that most broods break up between six and eight weeks after hatching. Large numbers that I have caught in summer fields from June 10 to September were not flying together in broods; there is a preponderance of juvenile males in these captures, whereas at hatching the sex ratio presumably is even, and in the fall sex ratios of juveniles from thousands of hunter wing collections show close to a 50:50 sex ratio.

Behavior of Hen with Brood

The mother woodcock is very solicitous of her offspring (Figure 28). A hen brooding her chicks can be caught with a hand net, so reluctant is she to move. When flushed from her brood she makes a labored flight with her legs hanging and her tail depressed (Figure

Figure 29. Decoy flight of hen

29). I have seen her lead a dog away, keeping just a jump ahead of him. When she has lured an enemy 100 yards or more from her chicks, she resumes rapid flight, rising swiftly. Then quickly circling around, she alights within a few feet of her brood. If the chicks are a week or more old and off the nest, she calls them with a low *chur-chur*.

The literature on the American and European Woodcock is replete with accounts of hen woodcock carrying their young. Mousely (1933) cites early records of such observations in the British Isles in the eighteenth century. Such reliable naturalists as Grinnell (1922), Schorger (1929), and Preble (1927) recorded instances of the American Woodcock hen transporting young. Schorger reported the hen carrying the chick between her "toes." Preble and Grinnell reported the woodcock carrying the chick between her thighs. Mendall and Aldous, Wright (letter), Liscinsky (1962), Blankenship and others who have observed and banded hundreds of broods, have never seen a hen carrying young, nor have I. It is possible that a hen woodcock suddenly flushed while brooding small chicks might accidentally catch a chick between its thighs and carry it a short distance. In 1950 I flushed a Ruffed Grouse hen from a nest where she was brooding a clutch of newly hatched chicks, one of which, purely by accident, was air-lifted 20 feet from the nest in this manner.

I have already described the peculiar labored flight of a hen flushed from her brood. Her deeply depressed tail and hanging feet could give one the impression she was carrying something. This, I believe, has given rise to many of the accounts of hens transporting young.

On the other hand, I have made one observation that still puzzles me. On May 15, 1952, while I was driving on a macadam road at midday, I saw a hen woodcock fly across ahead of the car. Her slow flight, drooping tail, and dangling legs indicated that she had flushed from a brood. She dropped into the woods 20 feet from the shoulder of the road. Parking a short distance beyond her crossing point, I returned with my bird dog. The woodcock again flushed with the labored flight that indicates the presence of young. Searching the ground, I found a single chick two days old. Had the hen been carrying it when she crossed the road? I still cannot be certain; it is unusual to find a single chick alone at such an early age. And I have never seen a hen woodcock *return* to a brood while using that slow, labored flight; typically the return flight to the chicks is swift and direct. The evidence was entirely circumstantial, but it does exist even though it is contrary to any other experience that I or other workers have had. I pass the incident on for what it may be worth. In the light of reports of this behavior, particularly from Europe (Chapter XI) I am not content to relegate the story that woodcock carry their young entirely to the realm of nature myth and folklore. In the hundreds of accounts of this behavior in both species of woodcock it is difficult to separate fable from fact. There should be no reason to doubt the records of reliable ornithologists, but I believe that whether the matter is fact or myth can be resolved only by collecting a hen carrying her young or by obtaining a photograph of the event.

Brood Cover

Mendall and Aldous state, "The cover where the broods are usually found does not differ materially from that used in nesting." They found 79 broods in cover of birch, aspen, spruce or maple, and 20 broods in alders. In Massachusetts I have found broods in all kinds of young mixed forest stands, in open fields, brush, road edges, alders, aspens, and birch. The hen leads her brood to poorly drained areas where the chicks can probe for worms or other invertebrates.

CHAPTER VI

Feeding Habits

The diverse methods of obtaining and handling food by birds are intriguing adaptations to their environment. Chief among these is the construction and form of the beaks of various species. Warblers and other small insectivorous birds depend on fine, thin forcep-like beaks for picking insects from vegetation. The bills of flycatchers and swallows are broadened sidewise and provided with bristle-like feathers for catching prey in the air. Finch bills are heavy and adapted for opening heavy seeds. A curious modification of the finch beak is that of the crossbill in which the tips of the lower and upper beak do not meet but cross. It serves as a wedge to pry scales off pine cones, thus exposing seeds which are deftly picked by its tips. Most people are familiar with the chisel-like design of a woodpecker's beak for prying apart decayed wood, or the hooked beak of the birds of prey for tearing flesh. There are some bizarre bill shapes in the avian world. In the New Zealand Wrybill Plover (*Anahynchus*) the bill tip angles to the right. The theory, according to Allen (1925), is that this plover feeds by following around bases of stones, keeping always to the right. Another New Zealand bird, almost extinct, the Huia (*Heteralocha*), is quite unique in that the bills of the two sexes are shaped differently. Huias are said to mate for life and invariably feed together. The male opens grub tunnels with his straight sharp beak and the female extracts grubs with her longer slender bill.

The woodcock depends on its long beak for extracting worms from the soil. As the woodcock, through evolution, developed its peculiar bill, its legs gradually shortened as it moved from the water and marshes to the uplands. Many of its wading relatives, such as the godwits and curlews, have long legs enabling them to stand and feed in several inches of water.

FOOD

Most analyses of woodcock stomachs were from birds collected in the fall or winter. There is little question that angle worms are the staple diet throughout the year, but I believe the woodcock is an opportunistic feeder. It will devour many other invertebrates when they are available, and it will occasionally feed on vegetation.

Sperry (1940) reported 90 per cent animal food in the stomachs of 261 woodcock collected in the United States and Canada. Earth worms made up 68 per cent of all food eaten. Sperry found the highest percentage of insects (38 per cent) in August stomachs. I collected 15 woodcock which were feeding on invertebrates other than worms on open dry fields in summer (Table N, Appendix II). Beetle larvae were the commonest food item of woodcock in summer fields. I have observed woodcock eating ants and catching moths in the air (when attracted by light). Aldous (1939) found earthworms made up 86 per cent of 63 woodcock gizzard contents collected in October in Maine; Pettingill analyzed 70 woodcock stomachs in October in Nova Scotia; they contained 86 per cent angleworms.

Vegetable items found by Sperry (1940) included seeds of sedges, violet, alder, raspberries, blackberries, cinquefoil, ragweed, bedstraw, smartweeds, elder, dogwoods, and pigeon grass. One woodcock stomach contained ferns exclusively. In these analyses vegetable food has rarely comprised more than 10 per cent of the content of the stomachs. A New Jersey state game warden, Joseph Mathis, maintained a feeding station for Bobwhites during a very cold spell in February and early March, 1934. Cracked corn was put out at Gretna, New Jersey, and he saw five woodcock feeding on the corn along with a covey of Bobwhites. The woodcock continued to feed on the corn for a month (Cottam, 1934). Miller (1957) analyzed 115 fallshot birds in Pennsylvania; by volumetric percentages there was 60 per cent plant debris, 17 per cent earthworms, and 14 per cent arthropods. Seeds of blackberry and sedges (*Carex*) appeared most frequently, and fly (*Diptera*) larvae were the commonest insects. Salamanders, and in one instance a frog, have been found in woodcock stomachs. Dexter (1949) found a woodcock stomach full of three species of aquatic snails.

I had the opportunity to observe captive woodcock over several years; from 1958 through 1964 I held groups of birds (a total of

162) for limited periods each summer. They were confined in pens of varying sizes. Their diet was worms, weighed and placed in pans countersunk in the ground; a trough of running water was provided. Each bird ate an average of 150 grams of worms each 24 hours. This approximated the average summer weight of my captives. Liscinsky (1962) fed two woodcock on half their weight for 12 days; each bird had lost 30 per cent of captive weight by the time it was released. It is not surprising that woodcock require such a high number of worms when one considers that 85 per cent of weight of earthworms is water (Grant, 1955). Such arthropods as centipedes, ants, wireworms and beetle larvae probably contain a higher percentage of protein in ratio to their bulk than earthworms. If these insects could be fed to captive woodcock, I believe the birds would maintain their weight by consuming less than their body weight each day. Glasgow (1958) and other investigators found that worms formed almost the exclusive diet on the wintering grounds. In view of the apparent dependence of woodcock on angleworms, I investigated the ecology and distribution of worms in North America.

After the passage of glaciers in the Northeast 10,000 years ago, there were no native worms in Maine (Gates, 1961). Today they are common in that state, which is an important part of the breeding range of the woodcock. The geographical range of woodcock after the passing of the Ice Age is not known. Gates (1961) states that the 22 species of earthworms found today in Maine are exotic and were introduced in one way or another by colonists in the seventeenth and eighteenth centuries. If this is true, the spread of worms must be rapid, since Smith (1919) found one species of worm (*Lumbricus variegatus*) to have an extensive range across northern regions of Europe, Asia, and North America. Woodcock reported in James Bay may feed on this species. There are only two native worm genera (*Bimastos* and *Eisenia longbeogi*) in Massachusetts (Stebbings 1961, personal communication). *Eisenia* is strictly a streambank form and *Bimastos* has never been found in numbers.

Murchie (1961, letter) has made some intensive studies of earthworms in Michigan. Although many forms are exotic, he believes there may well have been sufficient native worms to support woodcock in the central part of their range before the introduction of the exotic forms. Herman (1952) made the only careful study of earthworms in Louisiana and his research was confined to one parish. Of 12 species he found, four were introduced from Europe, three from the Orient, and five were native. There is one record of Indians eating woodcock in Quebec City in 1634 (Pettingill). Assuming the woodcock is now dependent on worms, it must have invaded its northern breeding grounds after the arrival of colonists. Possibly, the bird ate other types of food before earthworms were introduced.

Mendall and Aldous took over 600 samples of soil and earthworm counts in Maine. They found worms in soil with a pH between 5.00 and 5.75. Stegeman (1960) found that amount and diversity of humus materials rather than acidity in the soil is the determining factor in earthworm distribution. Fayle (1961) stated that deep-working worms (such as *Lumbricus terrestris*) do not tolerate acid conditions, whereas shallow-working worms thrive in acid soils. Stebbings (1961, personal communication), studying earthworms in the Harvard Forest in central Massachusetts, found good earthworm populations in hardwoods where the soil was damp. Although damp, these areas were not poorly drained. Due to subsoil conditions, drainage water came near the surface; there was constant water movement, resulting in aeration, which created better worm habitat.

I believe there is some correlation between tree species and earthworm distribution. Handley (1954) discussed nutrient properties of litter from different tree species in England. Several tree genera characteristic of woodcock habitat ranked high in nutrients. For example, litter under alders had by far the highest percentage of nitrogen (2.54 per cent) of any of the 24 tree genera listed. Alder is one of the very few plants which has the same attribute as legumes, in that it hosts nitrogen fixing bacteria in the soil. Aspen ranked fourth and birch eleventh. Several other fast-growing species such as black locust and tree of heaven, found in good woodcock habitat in certain areas, were not included. There may well be a relationship between nitrogen per cent in litter and earthworm abundance. Mendall and Aldous found a correlation between earthworm abundance and woodcock occurrence. The important point in their work was that they found the highest counts of earthworms in alder coverts or mixed coverts containing alders where one expects to find more worms and woodcock.

Liscinsky (1962) has made intensive studies in Pennsylvania of earthworm-woodcock relationships, finding a correlation between earthworm abundance and woodcock distribution. Miller (1957), in a Pennsylvania study, recorded soil properties, including drainage, texture, acidity, and organic matter, for a number of study plots in alder coverts. Although he found earthworms in all plots in greater or lesser number, he concluded there was no correlation between number of worms and number of woodcock. Liscinsky (letter) stated that this was a year of heavy precipitation and that many plots contained earthworms which would be absent under normal weather conditions. Glasgow (1958) made several similar studies on the wintering grounds and found that cover rather than abundance of worms determined woodcock distribution on feeding fields. In summary, there is a correlation between woodcock occurrence and earthworm numbers only if other requirements of woodcock coverts are met.

FEEDING BEHAVIOR ON BREEDING RANGE

Captive Woodcock

In one of my pens holding 20 to 30 woodcock in the summer of 1959, I set up an automatic camera with flash bulbs near the feeding pens. The film was exposed every time a bird passed a light beam. Each exposure photographed a clock and a thermometer. Over 200 photographs were exposed. I recognize that under these artificial conditions the birds' behavior may have been atypical, but I found it interesting. The birds fed at dawn or just before, at noon, and in late evening. More birds fed at noon than other periods even with temperatures as high as 95° F. Relatively little feeding occurred in the middle of the night. If one bird started to probe in the pens it appeared to stimulate foraging by several birds. Most exposures showed a group of birds feeding together.

I was impressed by the dependence of birds on water. They frequently drank; unlike many birds which hold their beaks vertically when swallowing water, the woodcock appeared to suck water. After probing, a bird would walk in the trough of running water and clean its bill with one of its feet.

Figure 30. Woodcock probing (Photo by Wilfred Kimber)

With an abundance of worms available in a concentrated area the captive woodcock quickly learned to locate the source of food, but every probe was not always successful, especially when the feeding pan was crowded. If a bird failed to obtain a worm in one attempt, he withdrew his beak quickly and probed rapidly until he caught one (Figure 30, 31). On occasion two birds would get different ends of the same worm and engage in a comical tug of war.

Wild Woodcock

On April 12, 1939, Mendall and Aldous had an unusual opportunity to observe a group of feeding woodcock in Hancock County, Maine: "Through the center of this cover ran a slow-flowing spring 'seep.' The ground was not frozen and between 30 and 40 woodcock were scattered along the seep, most of them feeding. The time was 10:00 A.M., the sky overcast with temperature above freezing. There were approximately two inches of new-fallen snow over which the birds were walking. At times they moved slowly, in such a manner as to give the impression they were carefully feeling their way. . . .

Figure 31. Woodcock feeding (Photo by Wilfred Kimber)

When probing they would insert the full length of the bill into the soft earth."

Glasgow (1958) has described vividly the feeding activity of three woodcock in mid-afternoon on a dry lawn in East Texas: "These woodcock walked a few steps in a rumba-like manner, sat down, placed their beaks on the ground and sometimes arose in a half-crouch and probed their beaks in the ground in a jackhammer fashion. They did not always probe when they sat down. Sometimes they walked a few feet or circled around and repeated the same performance."

I have observed woodcock feeding on small invertebrates in dry summer fields. They feed very rapidly on such surface insects as ants. I shot two woodcock in Vermont in flight at mid-morning with worms in their bills; one had been feeding in an aspen covert and the other under a group of pines.

Summer Flights

There is a dearth of information on cover and woodcock feeding activities during the critical summer months when the young are reared and molting takes place. Pettingill cites a number of references emphasizing the bird's desperate quest for food during the dry periods. He mentions reports of others who have observed woodcock during summer frequenting watered lawns, cornfields, pastures and briery thickets; Mendall and Aldous believe the birds seek thickets with wet ground. Blankenship found woodcock concentrating in summer near more permanent wet spots, such as stream bottoms. Liscinsky, in Pennsylvania, found most summer birds in bottomland coverts where alder, silky dogwood, crab apple, and hawthorn were the predominating plants.

Fifty-three of 100 summer flushes in my Massachusetts study areas were from alder thickets, 12 from stream bottoms, and 35 from conifer plantations. Shade provided by dense conifers is an integral feature of woodcock habitat in New England in hot weather. Woodcock use these areas for daytime resting, whereas the moister sites provide angleworms. I discovered that in summer twilight woodcock are very active, and more gregarious than at any season of the year, except winter, for I caught 761 woodcock with Japanese mist nets in summer evenings from 1952 to 1961.

I traced birds observed in flight in the evening if they were flying in a consistent direction, and found these woodcock alighting in certain fields. Aylward (letter) reported accompanying a market gunner during the last century when the Massachusetts woodcock season opened in July. He described dusk shooting of woodcock flying to the slope of a dry hill in Essex County.

I found four fields frequented at widely scattered locations in Quabbin Reservation. I used three of these fields as study and netting areas; the fourth was open to the public and unsuitable for effective netting. Evening flights of woodcock began after the breeding activities ceased. The earliest date a bird was captured was June 13 and the latest October 10. The heaviest evening flights occurred in July.

All the fields were bordered by trees or shrubs 20 feet high or less, but grass or shrub cover varied in each area. Whatever the cover, there were a few relatively clear patches on the ground. Woodcock like to be able to walk around easily, whether feeding or engaged in other activities. These fields were close to regions where a large number of singing males were heard in the spring, the size varying from one and one-half to ten acres. All were used as singing grounds in the mating season.

The smallest field was covered with low-bush blueberries, scattered clumps of oak and chestnut sprouts and a scattering of sweet fern. It was the site of an old burn and bordered by gray birch about 20 feet high. Rotting logs provided habitat for various types of arthropod larvae; ants were abundant. A three-acre field was relatively bare, having been scraped by a bulldozer ten years before; on this, the only food appeared to be ants. The larger fields were covered with grass clumps and bordered by pines and mixed forest. All these sites were dry, so woodcock fed from prey on the ground surface.

I aged all birds captured through the end of July by the color of the tips of their scapular and back feathers as described by Duvall (1956). August and September birds could not be aged with certainty during night banding until Martin (1964) discovered how to age the birds by the color pattern of the secondaries. Sex was determined by bill length and width of outer primaries (Greeley, 1953).

The sex and age groups in order of number of captures were juvenile males, adult females, juvenile females, and adult males (Figure 56 and Table O, Appendix II). I do not consider such age and sex differences representative of the summer population, but they may well reflect differences in behavior among the different categories.

Behavior during Flights

Woodcock were seen flying into the fields one-half hour after sunset on clear nights at the same light intensity as the beginning of the courtship performance earlier in the year (Figure 32). Within 5 to 15 minutes of the appearance of the first bird, no more birds flew into the field. On clear nights birds remained in the field from 10 to 30 minutes unless flushed. On damp dark nights they sometimes remained two or three hours.

They came in singly as well as in groups of two, three and four. Although Figure 32 shows birds approaching from all points of the compass, the majority came from the western sector and often circled the field and made the final approach from any direction. The pattern of approach depicted is typical of the flights of incoming birds in the other netting areas. Between 40 and 50 birds have been counted passing through or alighting in these fields the same evening.

Figure 32. Summer evening flight patterns

In July, 1955, Dr. William Nutting observed the antics of an apparent male and female on the netting ground at very short range.

> Heard peent NE at 8:52, and another weak one at 8:54 E.D.T. Several birds in from NNE. At 8:58, one bird flew in with wing whistle six and one-half feet from me. He stood and turned SW. Then, I noticed another bird walking in from SW. First bird moved to meet the new one. Bird 1 (hereafter "he") made aggressive head pass toward second ("she"). She stopped two feet from him. He raised his wings, lowered and advanced, raised wings again, lowered and moved; then, in several fluttery wing moves, came up to female and passed his bill at her mid-back. She then moved off three feet NNW. He moved two and one-half feet NE of her. Both stood still. I heard her give a soft cat wheeze note—followed in a few seconds by another. He turned around and squatted (faced from her). He rose, appeared to peck at ground, then turned facing her. Suddenly he flew in a low arc over her head, wings whistling, to

> land about twelve feet beyond her. She did not change position. In one minute, he flew up with wings whistling and did a semi-courtship flight, directly overhead, landing about fifty feet to SW. She walked slowly out of sight NW.
>
> I crept toward her, and flushed her—her wings fluttered, but didn't whistle. Time 9:14.

Nutting's description of the behavior of the bird presumed to be a male is similar to the actions of a breeding male preceding copulation.

Effect of Weather on Summer Activities

Windy nights, with an air movement of more than five miles per hour, usually yielded poor catches. The motions of the nets were often detected by the birds, and thus avoided. Still, hot, humid evenings following afternoon thunder showers created conditions which were most consistently productive of birds and often stimulated insect activities. Woodcock were also active on quiet evenings, with a light drizzle or fog. Clear nights with a full moon caused erratic behavior, and few birds were captured.

At dawn on July 18, 1959, 64 birds were counted on three miles of dirt road during a light rain. No birds were seen two mornings later over the same route under clear weather conditions. Ammann (letter) cites a record of a Michigan hunter who counted 48 woodcock along one mile of a dirt road on September 12, 1962, following a light rain; these may have been migrants.

Repeats

I wondered if the feeding flights were local movements by different groups of birds each evening. In netting Area 2, I found that 20 per cent of the captures in 1956 and 1958 were repeats. The chances of capturing repeats were small, partly because seldom more than 30 per cent of the birds observed each evening were netted. And, perhaps more importantly, because the flight pattern of individuals varies from evening to evening. Birds in netting Area 2 came from a relatively concentrated woodcock range. In contrast, Area 1, which yielded about 400 birds, produced few repeats. Birds from this area were surrounded by hundreds of acres of woodcock range, and some were taken as repeats and returns several miles away. Since the field is small, many other alighting areas must exist. Limited data on returns (see p. 111) suggest we were dealing with resident birds.

The distance individual birds flew in the evening to reach the feeding fields is somewhat conjectural. I posted observers at intervals near Area 1; they saw many birds flying east over a large hill half a mile fom the landing field. In Area 2, we saw many birds approaching the field from at least one-third of a mile. I never saw a bird rising from diurnal cover and flying to the field. In one instance, a graduate student captured two juvenile males in a funnel trap on the edge of a large alder swamp two miles from the netting field and separated from the field by a large hill. Ten days later these birds were netted, but their whereabouts between the two capture dates was unknown. Martin (personal communication) reported that birds netted in Maine summer feeding fields have been recaptured in funnel traps up to three and one-half miles from the netting site. Similar behavior has recently been observed in Michigan (Martin, 1964), so we may conclude that woodcock are more active in summer than heretofore reported.

Fall Behavior

Woodcock hunters have recorded instances of leaving a covert at dusk and watching birds alighting "like falling leaves" in open fields or alders. An unsportsmanlike practice indulged in by some is shooting woodcock against the fading evening sky after sunset when the birds fly to feeding areas. Such gunning is referred to as "dusking" woodcock. Numbers of woodcock, dropping into openings in fall, leave at dawn to scatter to diurnal coverts. Many sportsmen interpret this evening flight as migratory, since they fail to find birds there the following morning.

I witnessed an interesting evening movement of woodcock on the side of a Vermont mountain on a hot mid-October day in 1953. The site was a large abandoned farm which contained 40 to 50 acres of white birch, willow, aspen, crab apple, and hardhack. The last plant grew densely in the open fields. With two other hunters, accompanied by well-trained bird dogs, I spent the morning intensely hunting the covert. For the amount of fresh sign and cover, only a few woodcock were found. At midday the temperature approached 80°F. In the early afternoon, when we were leaving the covert, the dogs cast through some dense stands of red spruces in the lower perimeter of the covert. At once they made points, and I flushed several woodcock from the heavy shade under spruces so thick I could not shoot. I revisited the covert late in the afternoon of the same day. At sundown approximately ten birds flew at a low altitude from the spruces and dropped into the open fields of hard hack to feed. As it grew darker others were heard but not seen. These ten were conspicuous, but it is probable many others flew or walked from their shaded retreats to the aspen and birch covers where so much sign had been observed in the morning. Direct observations by one hunter, W. O. Clark, and the trailing behavior of bird dogs on occasion indicate that woodcock often will run or walk 200 or 300 yards. The daily ritual of concentrating on feeding areas at dusk and departing for scattered daytime resting places at dawn is probably as typical of woodcock in the fall as in other seasons.

FEEDING BEHAVIOR ON WINTER RANGE

Glasgow (1958) described two general types of winter habitat for woodcock. One consists of agricultural fields and bottomlands in the Southeast (Figure 33), the other of open upland fields and of various species of pine. The upland pine regions are bisected by many streams bordered by numerous hardwoods. He made detailed observations in Louisiana, where large concentrations of woodcock make flights at dusk into fields of various types to feed on earthworms. This phenomenon is similar to summer evening flights, but with certain important differences. Instead of the dry upland fields where the birds feed in summer on a variety of invertebrates, the wintering birds enter much larger damp or wet fields and feed almost exclusively on earthworms. Birds by the hundreds often fly into the feeding areas.

Earthworms are not the only factor influencing the distribution of woodcock on fields. After checking 300 plots, one foot square by 3 inches deep, in fields where woodcock fed and in adjacent fields they avoided, Eneminger (1954) found that earthworms were

Figure 33. Winter feeding field, Louisiana (Photo by John D. Newsom)

equally abundant in the fields where there were no woodcock. Cover of weeds or drainage ditches in a field controlled the choice of feeding sites used by woodcock (Chapter IX).

The manner in which woodcock approach feeding fields depends upon cover bordering the field. On nights when flights were heavy the birds appeared to come from all directions to fields that were surrounded by daytime cover. Three or four birds could be seen in the air at once. Most fields were bordered on only one side by brush and swampy cover. In one such field under observation several woodcock were seen to fly into the field parallel to the edge of the swamp. Woodcock fly into the fields at twilight approximately half an hour after sunset on clear nights, responding to the same light intensity in any season or latitude.

Distances

Little definite information was obtained on distances woodcock travel to feeding areas. Glasgow observed many birds approaching from the direction of heavily timbered areas, which were separated from the feeding site by about three or four miles of open farmland. Norris (1941) stated that fishermen and trappers reported seeing woodcock in the heart of the Atchafalaya Swamp. Deer hunters and duck hunters also saw woodcock flying at dusk down bayous toward feeding sites about eight to ten miles in the interior of the swamp.

The fields at Ravenswood and Redcross, Louisiana, were separated by about four miles of woods. Four birds banded at one or the other of these locations were netted on the other site, indicating a known movement of at least two miles from the middle of this swamp. Near Grosse Tete, Glasgow recaptured four birds the same evening they were banded; one of them was about four miles from the banding site, the others only about a mile away. Since these birds had been disturbed in banding, their travel may not represent normal movements.

Woodcock remained on the feeding sites until early morning. They fed actively from twilight until about 11:00 P.M., when they were found resting. Feeding was resumed before daylight early in the morning. Departures from fields were made over a much longer period of time than arrivals. Normally departure occurred a few minutes after dawn, but on cloudy or rainy days woodcock sometimes remained in the fields until 8:00 or 9:00 A.M. Birds remained longer in fields with tall cover than in those with short vegetation. Occasionally a few birds stayed all day in feeding areas that contained clumps of high weeds or shrubs.

Birds did not always remain in the field. On cold, bright nights several alighted in the fields, peented once or twice and immediately flew back over the swamp. This behavior was especially noticeable on nights when the moon was bright, humidity reduced, and the temperature low enough to result in a frost. They also deserted fields under high winds.

Winter Habitat

Glasgow's (1958) description of the behavior and distribution of woodcock on their wintering grounds is the most authoritative account available. In one experiment Glasgow divided a 40-acre field in which several hundred birds had been banded into five equal parts. Superficially the field appeared uniform, but most of the birds frequented two of the five sections. Careful study of this field and several others demonstrated that moisture and cover were the two most important factors determining uneven distribution of birds on the fields. The low areas usually held the most moisture and cover. Most of the fields were mowed, but some of the wettest areas could not be mowed and contained weeds which spoiled the hay. Most of the spots used by woodcock had protective cover from one to six feet high interspersed with small open areas and sufficient moisture to form good habitat for earthworms.

If a field has lain fallow for a few years, plant succession is very rapid in the humid Louisiana climate. After one year of abandonment briers invade the field and are followed by brush which forms thickets at the end of five years, thus changing potential nighttime feeding fields to cover frequented by woodcock for resting areas during daylight. Pastures, fallow fields, cotton, corn and sugar cane tracts are all used as night feeding sites for woodcock in Louisiana.

Bottomland pastures used by woodcock as feeding sites contained Bermuda, Dallas, headland, and sweet grasses. Prevalent weeds included Yankee weed, cocklebur, sumpweed, smartweed, and fleabane. There was often a scattering of berry bushes, rose bushes and small trees, including rough-leaved dogwood, hackberry, water locust, and red gum. Both corn and cotton fields were extensively used by woodcock after stalks had been cut and the litter trampled by livestock. Cornfields often supported a dense growth of Johnson grass, crab grass, foxtail, millet, and signal grass, sesbania, chocolate plant, tie vine, and cocklebur.

When ground vegetation was too dense to be attractive to woodcock, the birds fed in fallow fields. These fields supported a dense growth of weeds three to eight feet tall with an understory of grass. This vegetation was made up of such species as sesbania, cocklebur, sumpweed, smartweed, Yankee weed, giant ragweed, foxtail, crab grass and signal grass. Grazing and burning made such fields more suitable for feeding by woodcock.

Glasgow's account of the practice of burning to attract birds is of particular interest; he found woodcock concentrating by day on brush cover surrounding burns. Night hunters burned fallow fields, pastures, levees or forest land to create habitat attractive to woodcock. Glasgow describes a recently burned field as follows:

> A 3-acre fallow field that contained a dense cover of grass and weeds was burned about 1:00 P.M. at Sherburne on January 29,

> 1954. Prior to burning, woodcock were observed in adjacent fields, but none had been found in the dense cover of this small field. The fire was still smoking at 6:30 P.M. when woodcock began to alight in the field. After dark, I banded 12 woodcock and flushed 6 others from the burned area. The fire removed practically all the grassy ground cover but did not destroy the cockleburs. Sampling for earthworms indicated that there was no difference in the number present nor the depth at which they were located in the burned and adjacent unburned fields. I believe that removal of the dense cover permitted woodcock to probe easily for earthworms in the grass roots where the worms were concentrated. In a check I made on this burn about a week later, one sample plot a foot square showed 52 probe holes, while many other foot-square plots contained 25 to 30.

Some fields where Glasgow formerly netted numerous woodcock were ruined as woodcock fields by the construction of canals which drained the moisture from the surrounding ground. Much of the natural land between the Atchafalaya Swamp and the Mississippi River was planted in sugar cane. After the harvesting of the cane, and if bordered by suitable daytime woodcock cover, the birds use the plantation sites for feeding.

Although the largest number of woodcock concentrate in the bottomlands in Louisiana, another important woodcock habitat is the uplands of pine. Woodcock concentrate in numerous abandoned home sites with small fields not yet invaded by pine. As long as there is sufficient moisture, woodcock feed on the hilltops where there is a minimum of soil erosion and thus more fertile soil than on the hillsides. When the high cover dried out, the woodcock seek wet seeps on the slopes or branch bottoms and flats, and any open area in the pinelands such as fire lanes or hog and armadillo rootings.

Various species of bluestem grasses dominate the herbaceous cover of the Louisiana stands of long-leaf, short-leaf and loblolly pines. Switch grass and sedges cover the creek bottoms. Carpet grass and bitterweed occur on heavily pastured hilltops where woodcock feed. Other plants on pine hill woodcock feeding sites are goat weed, cane flower, catgut, common lespedeza, blueberry, winged sumac, St. Andrew's cross, and berry briars.

CHAPTER VII

Migration and Homing

Woodcock migrations have fascinated and sometimes puzzled ornithologists and sportsmen for many generations. Woodcock, flying relatively unnoticed at night, migrate up and down the eastern half of the continent. One day, the local coverts may be filled with birds. The next, there may be none. Some of our knowledge of woodcock movements comes from hunters who, while pursuing grouse, quail, or rabbits, stumble onto groups of migrant woodcock. Sporting magazines, state game commission journals, and ornithological literature contain scattered records.

MIGRATION ROUTES

Glasgow (1958) plotted possible woodcock migration routes, based on 175 band returns (Fig. 34). Although his map is useful in depicting migration, it is based on so few returns that it must be considered tentative. His findings, however, were confirmed further after I mapped 225 other returns from birds collected at least 10″ longitude or latitude from the point of banding. In either case, the sample is still meager considering that over 15,000 woodcock have been banded.

Glasgow (1958) divided the northern breeding grounds into three general geographic regions: (1) New England and adjoining Maritime Provinces; (2) Ohio, Pennsylvania, New York, Quebec, Michigan, and southeastern Ontario; (3) Minnesota, Wisconsin, and the remainder of Ontario. For discussion these three zones will be referred to as (1) the Atlantic, (2) the Central, and (3) the Western. Since woodcock are scattered in suitable breeding habitat in an east-west band of over 1500 miles, these tentative migration routes have innumerable tributaries.

Glasgow (1958) suggested that migrating woodcock may be attracted to areas with sufficient moisture to maintain earthworms near the surface. I believe this can be a factor, but geographical features such as mountain ranges and river valleys may be more important in channeling migratory routes. I observed one migration on Martha's Vineyard Island, Massachusetts, when woodcock were not only found in the black locust swales of the native birds, but all over the pine and oak uplands where no birds were found before or after this migration. Pettingill and others have described hundreds of woodcock stopping among the sand dunes on outer Cape May during fall migrations.

My observations in Massachusetts suggest that early woodcock migrations occur in all types of habitat from the valleys to the tops of the Berkshires. Late migrations appear to be shunted into the river valleys. On November 15, 1962, a heavy late flight came through Massachusetts. I found no birds in the higher uplands, but from reports and limited personal investigations most of this migration was in the Connecticut River Valley. They were moving fast in front of a cold wave from the north.

Atlantic Route

The Atlantic route is used chiefly by woodcock that breed in the Maritime Provinces and New England. Seven Nova Scotia birds, however, were originally banded in Louisiana and appear to cross northern New England in their southern migration to funnel into the Central route to Louisiana. Over 90 per cent of nearly 100 returns from New Brunswick, Maine, and Massachusetts have been recovered on the Atlantic coast or piedmont. One Massachusetts bird and two from Maine have been recovered headed northeast, two of which were recovered in the piedmont in Georgia and South Carolina. As Mendall and Aldous suggest, sight records and band returns indicate that woodcock from the interior of the Maritime Provinces and New England migrate to the Gulf of Maine, moving in a southwesterly direction in a broad belt along the Maine and New Hampshire coasts. Some move across interior Massachusetts to New Jersey and Delaware. There is a segment, however, which follows the coast closely, crossing Cape Cod, often stopping on Martha's Vineyard or the Elizabeth Islands, proceeding to Long Island, and thence

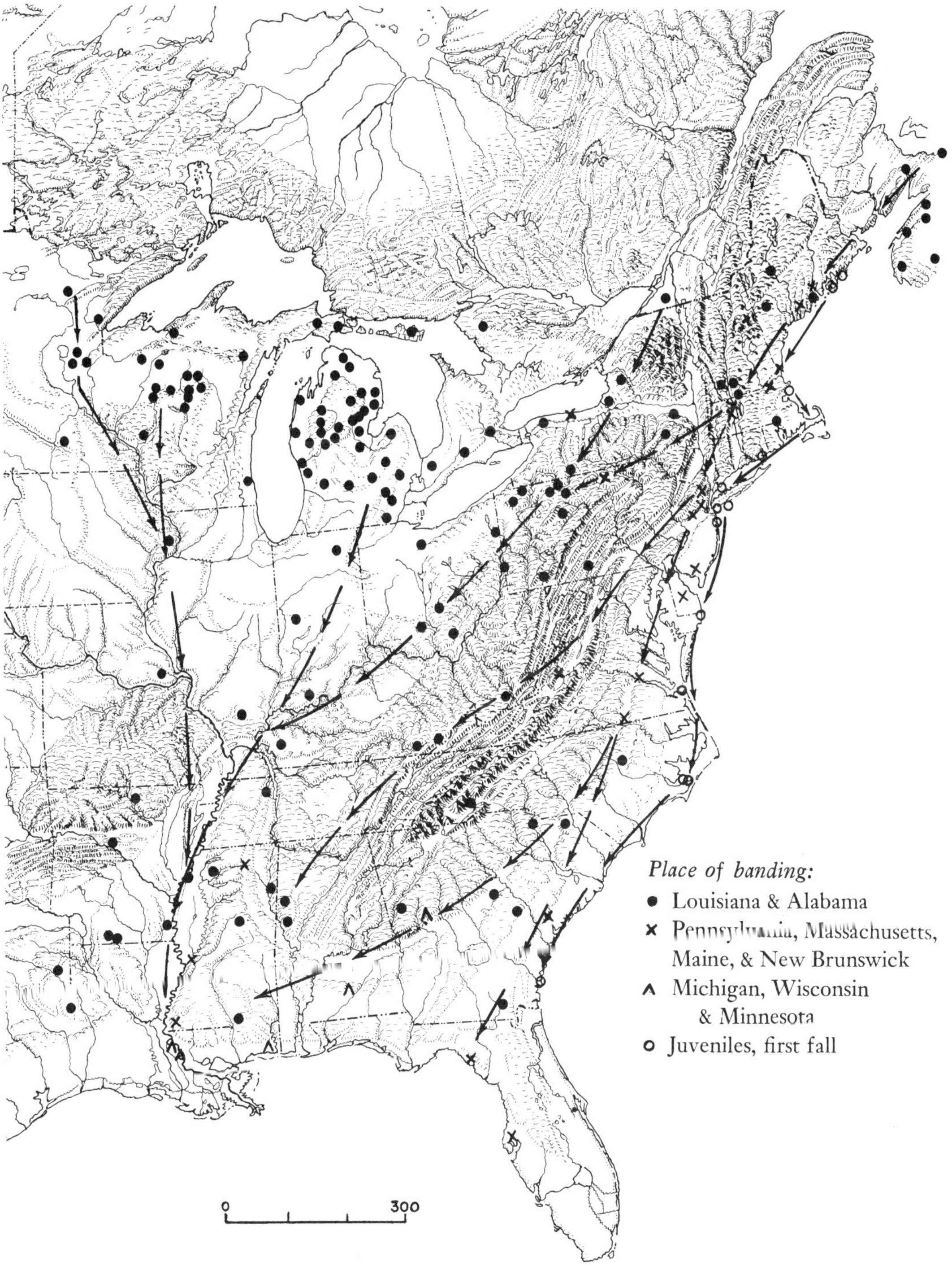

Figure 34. Migration routes

to Cape May where the flight lines divide. Some follow the interior route and some stay on the coast. Wing, age and sex analyses from wings suggest there is a high percentage of immatures among the coast migrants. Fragmentary band return analyses confirm this; of 19 recoveries on the Atlantic coast, 17 were banded in the previous spring or summer as juveniles. Birds crossing the interior of Massachusetts probably join segments migrating down the Connecticut Valley as well as some from the Champlain-Hudson valleys.

From northern New Jersey, woodcock spread out over the Atlantic coastal plain and piedmont, then proceed southward across Delaware, Virginia, North Carolina, South Carolina, Georgia, and Florida. The distance these woodcock move south may be dictated by the severity of the weather. There is some evidence to indicate that a few birds may follow a wide sweeping curve around the southern tip of the Appalachian Mountains to Louisiana. Woodcock traveling southwest by west out of New England travel a straight course, and cross southeastern New York to northwestern Pennsylvania, where they meet birds moving southwest just south of Lake Huron and Lake Erie. The direct return of a woodcock banded in Maine and recovered across the mountains in northern Pennsylvania indicates that some birds may follow this route. More recent band returns suggest that substantial numbers of woodcock leave the Atlantic route near this point in central New England and continue southward in the Central route.

The Great Valley, lying between the Blue Ridge crest on the east and the Appalachian Ridge on the west, extends from the Canadian border in eastern New York to central Alabama and offers woodcock a direct route toward Louisiana from the northeast. It is one of the world's longest valleys and is made up of segments of different river valleys, including the Champlain-Hudson, the Lebanon or northwestern New Jersey and southeastern New York, the Lehigh in southeastern Pennsylvania, the Shenandoah in western Virginia, the Clinch and Holsten in southwestern Virginia, and the Tennessee Valley in eastern Tennessee. According to White and Foscue (1943), in most sections this valley is broad and relatively flat, but in Pennsylvania and southwestern Virginia it is made up of a series of parallel ridges and valleys extending in a northeast-southwest direction. There are no barriers of any magnitude to woodcock migration in this valley. The soils are fertile; fall precipitation is greater than in surrounding regions; cover is moist, hillsides seepy and slopes are wooded.

Three woodcock bands were returned from deep in this valley or from sites close to it, and three additional bands were returned from near its southern extremity. Access to the valley above southern Virginia is easy for woodcock moving southwestward out of New England. In fact, birds penetrating far inland in the area between Massachusetts and New Jersey either enter or cross this corridor. Because of persistent reports of scarcity of woodcock in southeastern Pennsylvania, I believe that most woodcock using the valley enter it in Virginia. Data needed to substantiate the existence of this flight route, however, are rather meager.

Central Route

The Central route is probably traveled by more woodcock than any other travel zone. This route, with its southern terminus in Louisiana and Mississippi, extends northward across west-central Tennessee and western Kentucky; swings northeastward parallel to the Ohio River through northern Kentucky, southern Indiana, southern Ohio, and continues through northwestern Pennsylvania and western New York to Quebec. Liscinsky (1962) states that for the past five years the greatest migratory flights of woodcock in Pennsylvania have been reported from the northwestern section of the state. The band returns plotted in Figure 34 show the continuity of this route.

The crossover between the Atlantic and Central routes through northern Pennsylvania has been described previously. Major northward tributaries exist along that section that lies between southeastern Illinois and south-central Ohio. Birds that breed in Michigan and southern Ontario, as well as those from Illinois, Indiana and Ohio, use these tributary routes to enter the Central zone. Although few bands have been returned from these central states, probably nearly all Michigan woodcock cross them. Illinois seldom has an open season on woodcock; therefore few bands are reported from the state. Pitelka (1943) reported woodcock as common spring migrants in the Champagne-Urbana region, and during the 1870's many woodcock were shipped to the New York market from the Kankakee region of

Illinois and Indiana. From hunter kill records obtained by mail surveys over a 14-year period, Barnes (1955) has shown that migratory woodcock are widely distributed throughout Indiana, with the peak annual kill shifting rather randomly from one region to another. W. C. Gilfillan (letter) wrote that, with the exception of a group of breeding woodcock in northeastern Ohio, few birds were reported in that state. Earlier records indicated that woodcock were regular migrants across the northwestern part of Ohio.

Since southern Michigan is approximately 250 air miles from the Ohio River, woodcock could easily fly this distance nonstop in a single night. But it is doubtful that they do, for Ohio, Indiana and Illinois have moist, wooded cover that provides good habitat. Since woodcock are moved by cold weather, it is possible that low temperatures moving down from the north extend over the three-state region about the same time each year and force woodcock south rapidly. Therefore, they may be present only briefly in this region.

Western Route

The Western route extends from Louisiana and eastern Texas almost due north to Wisconsin and Minnesota. Woodcock that follow this route fan out across Arkansas and southern Missouri, then follow the Mississippi River to within a short distance of their destinations before branching off to upper central Wisconsin and east-central Minnesota. In the southern third of this zone considerable mixing of birds from the Central route probably occurs. Some birds apparently branch out from the Western route and move in a northwesterly direction up the major drainage basins. Their western movement is undoubtedly halted by scarcity of moisture and a consequent lack of necessary food and cover.

It is apparent that the reader may, in a few instances, be required to use his imagination to see the continuity of routes that has been pointed out. Therefore, it is useless, without additional data, to delineate further subdivisions. With the return of more bands it may be necessary to discard some of the routes postulated above, or to describe new ones. Because of low hunting pressure, and the presence of woodcock in the central part of the United States for only a short period of time, there is a hiatus in our knowledge of migration, which can be eliminated only by continued banding.

SPRING MIGRATION

The severity of weather during January and February governs, to a large extent, the time when most of the wintering birds begin their northward migration. From 1952 to 1957, 8.0, 0.6, 0.0, 40.8, 16.2 and 8.8 per cent respectively of the total woodcock observed in Louisiana were found in February (Glasgow). February 1955 was relatively cold and damp, and woodcock remained until slightly past the middle of the month. During the 1954–55 season 41 per cent of all woodcock records were for February. During the eight-year banding period in Louisiana 14 per cent of the woodcock were caught in February. Because January 1954 was very warm, practically all woodcock had departed by the end of the month. In mild winters most birds had abandoned the vicinity of Baton Rouge by February 1, in normal winters by February 12, and in cold winters by February 29. Each year, however, a few birds were always reported in late February and early March.

Woodcock are exceptionally early migrants. There are many instances of the early arrival of woodcock in the northern breeding grounds even before the accumulation of snow has melted. It is not unusual for them to encounter spring storms during which they suffer from cold and snow.

In mild winters, some woodcock stay far north of Louisiana. It is generally accepted that most birds winter south of Missouri and Maryland, with only stragglers remaining above this latitude. Determination of migration dates below this latitude is difficult because resident as well as migratory woodcock are present during the winter. I. Grant Scott (letter) recorded a large number of woodcock wintering on Cape May, New Jersey, and A. Keniston (letter) reported that some woodcock winter on Martha's Vineyard Island.

Woodcock probably spend four to six weeks on their northward journey following a zone in which the ground has recently thawed. Although many records are available of early arrivals on the breeding ground, reports of dates on which the majority of the woodcock arrived are rather scarce. Russell (1954) reported what was believed to be a migratory flight of woodcock in Edmonson County, Kentucky, on March 12, 1954. In answers to a questionnaire, biologists in Kentucky believed the heaviest migratory spring flights occurred in March. Twitle (1890) wrote that woodcock arrive in northeastern Illinois between March 25 and April 30. Pitelka (1943) wrote that woodcock generally arrive in central Illinois during the first half of March.

Audubon (1840) reported heavy flights along the Ohio and Mississippi rivers from the middle of March to the middle of April in the early nineteenth century. Butler (1898) gave March 2 to 20 as the arrival period for woodcock in northern Indiana. Gilfillan (personal communication) reported that woodcock arrived in Ohio in late February and early March with heavy flights occurring in early April, but Dawson (1903) stated that most birds arrived in Ohio in March. Barrows (1912) wrote that the greater part of the woodcock population arrived in Michigan by April 1. Blankenship (1957) gave April 1 to 15 as dates when most woodcock are seen at Lansing, Michigan, and slightly later dates for northern Michigan. Ammann (personal communication) recorded peaks of spring migration near East Lansing, Michigan, on March 31, 1963, and two peaks in 1964, March 28 and April 7. Roberts (1936) reported April 8 as the average date for the arrival of woodcock in Minnesota, and Dangler and Marshall (1950) believed the birds arrive near Duluth, Minnesota, in April.

Liscinsky (1956) indicated that the bulk of the woodcock have arrived in Pennsylvania by the end of March. Eaton (1910) reported that the usual spring migration reaches southern New York about March 10 to 15 and the interior of the state two weeks later. C. P. Brown (1953), after conducting woodcock censuses in New York, believed most of the woodcock arrive in southern and central New York in late March and early April and in northern New York the first half of April. Mendall and Aldous (1943) wrote that woodcock become common in the northeastern part of their range in the latter part of March and early April. On March 18, 1952, I saw large numbers of woodcock at dusk on the Skyline Drive in Virginia. In March 1927 I found at least a dozen woodcock in a pasture within sound of breakers in Kennebunkport, Maine.

Spring Migration in Massachusetts

My observations and records for thirteen years in central Massachusetts indicate that the spring migration is far more predictable and less variable than the fall migration, regardless of weather conditions. Woodcock arrived late in the spring of 1956, when 44 inches of snow fell between the middle and end of March. In most years woodcock reached their breeding grounds the last week in March and first week of April. The Connecticut River Valley is apparently a migratory flyway; it is situated 800 feet nearer sea level than the study areas. Birds usually appeared in the valley during the first week of March. The earliest record of arrival in central Massachusetts was February 24.

The literature is replete with early migration dates of woodcock in various parts of its breeding range, but I am concerned here with the arrival dates of the bulk of the resident breeding populations. Resident adult males arrive in the vanguard of early migrants in Massachusetts study areas. There is no positive evidence that females arrive later than males, in spite of the fact that Glasgow (1958) found the females leaving their wintering range later than males. I took a few females in my singing-ground traps in the last week of March and found one complete clutch of eggs on March 27.

Rate of Migration

Wright (personal communication) reported on the rate of advance of spring migrants as follows:

> The date of arrival of woodcock on the Fredericton study area in spring was assumed to be the night before the first peenting bird was heard. The area was observed each night from March 28 until the first singing male was found. The same date was recorded at the Moosehorn Refuge near Calais, Maine. The distance between these two areas is 72 air miles. The dates of arrival at Greenbush, Maine, were recorded in 1953 by Westfall, and in 1954 and 1955 by Weeden. This area is 107 air miles southwest of Fredericton. These data were combined to give nine years' observations on the dates of first arrival in New Brunswick.
>
> The date of arrival at Fredericton 1953–1961 varied from April 1 to 12. The rate of advance of first arrivals varied from a low of 6.5 miles per day to a high of 53.6 miles per day. Measurements of wind, precipitation, pressure, and humidity were taken at 2:00 P.M. and 8:00 P.M. the day before arrival, and at 2:00 A.M. and 8:00 A.M. on the morning the birds reached Fredericton in the last eight

> years. These measurements showed that in four of the eight years there was some precipitation that night. Thirty-two wind recordings showed only two from east of the north-south axis. The highest velocity recorded was NW 20 with gusts to 38. Atmospheric pressure was always below normal and rising, but there appeared to be no definite correlation with frontal activity.

The spring return of bands from woodcock banded in Louisiana supports much of the above information. But the date on which a band was recovered does not necessarily represent the date of arrival of the bird at the recovery point, for it may have been in the area for some time. Of nine woodcock bands returned in the spring from Minnesota, Wisconsin, and Michigan, two were reported in the third week of March, six in the first three weeks of April, and one the first of May. Of five bands reported in the spring from Iowa, Illinois, Indiana, and Ohio, three were returned in March and two in April. One woodcock banded in Louisiana on January 5, was found dead on March 3 of the same year in eastern Massachusetts.

FALL MIGRATION

Many ornithologists have reported fall flights of woodcock following the first heavy frost in the North. Mendall and Aldous state:

> Fall flights appear to be more irregular and even more dependent on weather conditions than those of spring. In fact, because of the frequent and sometimes marked weather changes that often occur in September and October in the Northeast, it is difficult to prepare time tables for woodcock migrations, even when based on data accumulated for a number of years.
>
> It is believed that little definite migration starts, even in the northern limits of the range, until September. By far the bulk of the migrating birds pass through the northeastern region in the month of October, and during warm autumns in Maine many heavy flights have been recorded even as late as the first half of November.

Pettingill's data show that the time of concentration of heavy flights at Cape May, New Jersey, varied from late October to mid-December. Wildlife biologists in North Carolina think woodcock arrive there in the fall from October to December, depending on the weather to the north of the state. Griscom (1949) reported heavy flights about Concord, New Hampshire, for various years on October 19–20, October 15, and October 10–27. Bagg (1923) stated that woodcock flights arrived in the Connecticut River Valley of Massachusetts about the middle of October.

Brown (1953) presented peak fall migration dates of October 15–20 for New York State, and Eaton (1910) wrote that migration in New York State began after the first sharp frost in October, which was usually between October 1 and 15. Liscinsky (1956) wrote that migration began in Pennsylvania in October and continued to the third week of November. English (letter) believed that migration in Pennsylvania is practically complete by November 15. Bennett and English (1947) reported heavy flights of woodcock in northwestern Pennsylvania October 2–4 and October 10–11, 1946. Flights, believed to be the same ones, were observed in Center County on October 6–11 and October 15–19. W. C. Gilfillan (letter) wrote that in the fall woodcock drifted through Ohio during most of October, but that in mild seasons they were present until the first week of November. Under normal conditions the heaviest flights occurred between October 10 and 20. Severe cold snaps caused an influx of birds during October. Ammann (personal communication) reports most woodcock have departed from the Upper Peninsula of Michigan by October 22 and the Lower Peninsula by November 9. The average of nineteen fall migration dates for Ohio is November 7, and the average of four for Indiana is also November 7. The average for Kentucky is November 22; for Tennessee, November 13. Musselman (1945) reported a large concentration of woodcock feeding in central western Illinois November 13 to 26.

It is probable that the fall migration is influenced by freezing temperature, since most woodcock gradually move from the latitude of Kentucky and Tennessee and arrive in Louisiana during the latter part of November and early December. Forty thousand wings sent to the U. S. Fish and Wildlife Service from 1959–1962 give clues as to certain periods of heavy fall migrations during four years. Most of these data are broken down into the number of wings sent in by the week. Such factors as the opening date of the hunting season, weekends and weather influence the results. Study of the daily bags, weather records and dates might reveal further data on migration.

The percentage of the season's take by weeks in important states is presented in Table 1.

TABLE 1. Dates of Heaviest Woodcock Kill 1959–62

Opening Date Hunting Season	*State or Province*	*Dates*	*Percentage of 4-Year Kill*	*Number of Wings*
* October 1	New Brunswick	October 11–20	42	2,769
October 1	Maine	October 15–21	25	7,756
October 1	New Hampshire	October 12–18	24	2,769
October 5	New York	October 12–18	27	4,385
October 12	Pennsylvania	October 12–18	45	3,602
October 15	New Jersey	October 22–28	28	3,146
October 1	Michigan, central	October 1– 7	30	4,515
October 1	Michigan, northern	October 1– 7	39	1,350
October 1	Wisconsin	October 15-21	25	4,114

* New Brunswick season opens September 15, but for comparative purposes, I counted wings taken only after October 1.

I. Grant Scott (letter) reported that very heavy northwest winds and cold nights presaged large flights arriving on Cape May, New Jersey, and many observers in the North have recorded a large fall exodus of woodcock coincident with snow and freezing weather. Many woodcock arrive in the spring on their feeding grounds when snow covers much of the ground, whereas in the fall, snow prompts them to depart. This may appear as paradoxical behavior. I suspect that with the development of gonads in the spring, woodcock are highly stimulated to seek their breeding grounds during the increasing daylight. Although they often start nests while there are still subfreezing temperatures at night, the sun melts snow in patches of poorly-drained soil where woodcock can easily find food.

Fall Migration in New England

"The flight is in!" Among New England sportsmen such news is usually followed by frantic activity, extended weekends, and lots of spent gun shot. Weather conditions in the fall often prompt heavy southward migrations, but in many mild falls the dates of "flights" are totally unpredictable. There is no more controversial subject than what actually constitutes a flight. There are well-known Atlantic Coast points, such as Cape May, New Jersey, where the arrival of a flight is easily recognized; often several hundred birds will be found resting among the bushes just behind the beaches.

Glasgow (1958) finds the arrival dates of woodcock in concentrated wintering areas in Louisiana to be fairly consistent from year to year. Birds from New Brunswick and northern Maine often congregate in conspicuous concentrations in coverts on the southern Atlantic Coast. My observations in Massachusetts and Vermont suggest that, throughout the fall, migrating woodcock drift south in loose flocks, one group followed by another until late in the fall. There may be a group of migrating birds in one covert on a certain date, with none in a favorable covert only a few miles away. Hunters visiting the first covert report a "flight," and often assume that migrants in one covert indicate that a broad wave of woodcock has moved in from the north.

Much of the earlier fall migration may resemble the migration of robins, where large groups are often seen at different dates in different areas throughout the fall. In most cases, woodcock fly south in a leisurely fashion, stopping two or three days in one latitude. Woodcock lose only ten per cent more or less of their peak fall weight by the time they reach their wintering grounds (Chapter IV) in contrast to the non-stop flights of some shorebirds.

Instances of obvious flights include the following: On October 19, 1952, "hundreds" of woodcock were reported about 2:00 A.M. along a road on Martha's Vineyard. The hunting season opened the next day and quail hunters, rabbit hunters, and pheasant hunters reported finding woodcock everywhere, all over the uplands as well as in swales. These birds had crossed Cape Cod or stopped undetected on Cape Cod, and flown across Buzzard's Bay to the island. They departed two nights later and presumably had a long flight over open water before touching land again.

In 1955 a New Hampshire hunter reported a heavy flight of woodcock in the foothills of Cannon Mountain in northern New Hampshire on November 10, when the top of the mountain was covered with snow.

In 1961 there was a dearth of woodcock in central Massachusetts from October 20 to November 15. An early flight through the western part of the state had been reported in mid-October. By mid-November most woodcock hunters had assumed that all the birds had migrated, but pheasant hunters stumbled into a heavy flight in the Connecticut River Valley on November 15. One observer reported seeing "scores" just above the banks of the river. I went out after the peak of this flight and found some birds where none had been found during periodic checks earlier in the season. It is probable that such late flights travel more rapidly en route to their wintering grounds. One obvious way of recognizing a "flight" is to visit the same covert several times in a fall. If there are no birds one day, and a large group another day, there is good circumstantial evidence that a flight has come in.

Altitude and Flocking

On November 7, 1964, in Millbury, Worcester County, central Massachusetts, J. F. Le Blanc, a student at the University of Massachusetts, made the only direct observation of migration recorded during our studies. The woodcock Le Blanc observed began to appear at treetop level "shortly after sunset" when it was still light enough to make accurate counts. From a surrounding area of short poplars and alder, the site of an old piggery, he and his companion counted a minimum of 200 birds flying southeast. Of 15 "flocks," 13 were loose groups of "4 to 10 birds," but there were two closely bunched flocks of "30 to 50 birds." These passing flocks were seen over a period of 20 minutes until it became too dark to count; then other birds were heard but not seen.

Mendall and Aldous recorded two observations by others of migration. E. P. King, a former game warden, "states that at dusk one evening in late October, about 1920, in north central Maine, he saw a flock numbering thousands of birds." Mendall and Aldous also cite an observation by Louis Becket who reported seeing a loosely scattered flock of "several hundred birds" in the fall in eastern Maine as recently as 1930. These limited records suggest that woodcock migrate at low altitudes in loosely scattered flocks.

Departure of "Native" Birds

The notion that "native" birds leave first and are followed by "flight" birds is not necessarily true. After the Martha's Vineyard flight described above, "native" woodcock were found in abundance in the black locust swales (where chicks and adults had been banded in spring) until the end of November. I received six returns of resident males banded in the spring and shot in the vicinity of the banding site in late October or November after "flights" had gone through. Two banded birds were shot in late October, three on November 1, one on November 13, and one on November 17.

Arrival in Louisiana

Thorp (1846), in writing of "fire hunting," stated that "we presume these very birds return for their winter residence to Louisiana in the very months when fire-hunting takes place, which is the latter part of December, January and the first of February." Merovka (1939) stated that woodcock began to arrive about the middle of November and were most abundant from the middle of December to the end of January. Norris (1941) wrote that woodcock have been reported in Louisiana in numbers as early as October, but seemingly very few reach there before the middle of November and the main flight usually arrives during the first half of December. Lowery (1955) shows graphically that the greater part of the Louisiana wintering population is migratory, arriving in November and remaining until late January or early February. Records obtained by Glasgow (1958) confirm the findings of Lowery and Norris. Migratory woodcock begin to trickle into Louisiana in October and gradually increase in number reaching a peak about December 1. About 98.4 per cent of the birds banded in Louisiana are taken after November. These arrival dates vary somewhat with weather conditions to the north of Louisiana, but seldom by more than a week or ten days.

Winter Population (Glasgow)

After the wintering woodcock population in Louisiana builds up to a peak in December, it remains at a high level until the latter part of January or, in severe winters, to the middle of February. Of all woodcock banded by personnel of Louisiana State University and of the Louisiana Wildlife and Fisheries Commission, 42 per cent (3,280) were banded in December and 42 per cent (3,322) in January. From these figures it would be assumed that there was little difference in the woodcock population for two months. But woodcock were hunted on fewer nights in December, which may indicate that December is slightly better for banding work than January. From 1949–50 to 1956–57, of the total number of woodcock observed, 8,296 birds were observed in December and 6,985 in January. In five of the eight years more birds were seen per man-hour of hunting in December than in January.

Reid (1955) found more woodcock in the pine area of west central Louisiana in January and February than in December. In five out of six years, he recorded the greatest number in the field between January 15 and 29. In one year the highest number was found on February 5. This information seems to indicate that woodcock have started their spring migration, but it may be that they tend to desert the swampy areas in the Atchafalaya and Mississippi basins because of flooding and move to the upland areas where degree of soil moisture is more favorable.

HOMING

Banding studies offer abundant evidence that, as a rule rather than as an exception, migratory birds habitually return each year to the same breeding ground and wintering ground. As is well known, hole nesting passerine birds often return to the same nest box that they used the previous year; the woodcock is no exception in its homing ability.

Adult Males

When I trapped my first woodcock on his singing ground in 1949, I realized that large-scale trapping over several years would yield information on the survival and homing of adult males. Seven years of trapping produced 126 returns; since many returned to their breeding ground more than one year in succession, these records involved 84 individual birds. Most males returned in succeeding years to the close vicinity of their original breeding sites. Twenty-seven birds (32 per cent) were recaptured in precisely the same singing field where they were first trapped. One of them was trapped on a small isolated singing site five years in succession.

Since most males used more than one singing ground a year, I was not surprised to capture returning birds in fields other than those where they had been banded. Eighty per cent (68 birds) were taken within 1.0 miles of their first capture site (Table 2). Fifty-nine per cent (49 birds) were each caught on only two singing grounds.

TABLE 2. Distance Between Singing Grounds Occupied in Successive Years by 84 Male Woodcock on Return from Migration

Distance in Miles	*Number of Birds*	*Percentage*
0	27	32
0.1 –0.25	11	13
0.26–0.5	15	18
0.6 –1.0	15	18
1.1 –1.5	6	8
1.6 –2.0	3	3
2.1 –2.5	3	3
over 2.5	4	5

In addition to the trapped birds, 11 adult males were netted as returns on summer feeding fields one to five years after banding. Seven of these were retaken on a singing field in a subsequent spring and the remainder in a summer feeding field. None was recaptured more than two miles from its first capture site (Table O, Appendix II).

Adult Females

Only 17 females were trapped on singing grounds, so there was little chance of capturing a return in the spring. Eleven returns were netted in summer feeding fields; of these, eight were taken in the same field as the initial capture. One was recaptured in a field three miles from the first capture site. Time between captures varied from one to four years (Table O, Appendix II).

Juvenile Males

One juvenile male banded in its nest as a chick was captured on a singing field the following spring 150 yards from its banding site, and Mendall and Aldous collected a male as it descended from a flight song 400 yards from the nest where it had been banded as a chick the previous year. Of six juvenile male returns banded on the summer feeding fields, one was recaptured a year later on a singing field one half mile away, and the others were netted on summer fields. Of the latter, three were taken on the first capture site, one was netted one and a quarter mile distant, and the other three miles from its banding site (Table O, Appendix II).

Juvenile Females

One female banded as a hatchling was recaptured in a singing-ground trap two years later 300 yards from her banding site. I netted four returning juvenile females on the summer feeding fields one to three years after their first capture (Table O, Appendix II). Three of these were netted in the same field and another in a field ten miles away. A female banded as a chick by Mendall and Aldous in June 1940 was shot in October 1941 100 yards from its banding site.

In my studies, I found two possible exceptions to homing by females in hunter recoveries of birds I had banded in the summer. One juvenile banded on July 16, 1956, was shot October 1, 1957, in Peru, Vermont, 65 miles northwest of its banding site. Another female of undetermined age, which I had banded on a summer feeding field on September 1, 1956, was shot October 10, 1957, in Belfast, Maine, 350 miles northeast of the banding location. Since woodcock are known to begin migratory movements in September, this bird may have been a migrant when banded.

WINTER RETURNS

A few of the woodcock that Merovka (1939) and McIlhenny (1940b) banded in Louisiana were shot in subsequent years near their banding sites. Glasgow (1958) took 161 returns in Louisiana (Table Q, Appendix II). He believed that the majority had migrated to their northern breeding grounds in the spring and returned the following fall. He banded a woodcock on February 9, 1951, and recaptured it the following winter at its banding site; it was finally shot at Cumberland, Kentucky, on November 20, 1955.

Glasgow's records show a return of 2.4 per cent for all woodcock that he banded prior to the winter of 1956-57. It should be remembered that this figure is a minimum and does not represent the actual return, because many other woodcock that were not captured may have homed to areas near the banding sites.

Sex Ratio and Rate of Recovery

As shown in Table Q, 57 male and 71 female birds were recovered in Louisiana, a sex ratio of 80 males: 100 females. No males were recovered after the second year; six females were recovered three to five years after banding.

Distances

Seventy-one per cent of 161 woodcock recaptured in Louisiana were in their original banding fields or in adjoining fields. Among the unusual winter returns was one bird which was banded on January 3, 1951, recaptured December 19, 1951, and caught again on December 19, 1952, all on the same site. A male woodcock banded on December 31, 1951, was caught at its original banding location on January 29, 1952, December 31, 1952, November 29, 1953, and January 10, 1954. A female banded on January 27, 1952, was recaptured in the same field on February 21, 1955, and December 18, 1955—a span of four years.

In conclusion, I believe most woodcock of each age and sex group return to their original rearing grounds to breed and migrate to the same wintering grounds each year. It is highly probable, in spite of little evidence, that after its first year a woodcock retraces its original migration route.

CHAPTER VIII

Hunting

Sharp steel swished through the hay on the Perkins Farm. Each sure stroke of the scythe lengthened a windrow of rich timothy that would keep the cows through the long Vermont winter. Beyond the stone wall a fat crop of summer potatoes awaited the harvester's hoe. The apple trees behind the barn strained under the weight of ripening fruit.

Jonathan Perkins paused to mop the sweat from his face. It was a good year, 1850! The whole valley wore a mantle of prosperity. Below him the spire of the church in Bridgewater pushed skyward above the maples. To the right and left, the sturdy homes of his neighbors, like his own, glowed with fresh paint. Jonathan resumed his haying. This would be something to pass on to his sons and their sons and to future generations of Perkinses forever.

But change came to the valley and to the hills that Perkins loved. Each of five thousand rains robbed something from the soil. Granite ribs began to show through the thin skin of topsoil in the pasture and potato field. The cowpaths on the hillsides became ugly scars that deepened after every shower.

Jonathan Perkins, stubborn and independent, stuck it out to the end. But after the old man died, his sons took jobs in town. Birches, aspens, willow, pine, and hardhack invaded the pastures; the orchard went into disrepair. The last of Perkins' heirs who tried to work the soil quit in 1920. Time, wind, weather, and dry rot did the rest. Today only a cellar hole, tumbled stone walls, and a few gnarled apple trees in overgrown fields remain as monuments to Jonathan Perkins' dream.

In 1955 shadows of the Green Mountains stole across the old Perkins farm as, beyond the valley, the sun dipped below the ridge.

Figure 35. After woodcock (Etching by A. Lassell Ripley)

There was no wind. The crisp October air carried a foretaste of heavy frost. Occasionally one of the birches shed a few leaves that spiraled earthward like fragments of dying sunlight.

Two hunters worked a long finger of birches and aspens on the hill above the cellar hole. Each covered an opposite side of the run. Between and slightly ahead, a setter and a pointer cast through the covert. Their "birdy" tails indicated hot scent. Suddenly the setter stopped. He took two tentative steps, and froze. The young pointer, head high, backed on command.

The hunter on the left, gun ready, moved in for the flush. The other covered the opening to the right. For a few seconds the tableau remained unchanged. For woodcock hunters this is the "moment of truth." The staunch dogs are statues half screened by the alder stems. Thumbs tremble against safety catches. Scalps prickle, and short hairs bristle above corduroy coat collars. Hearts beat faster, and mouths become strangely dry.

The picture shattered in a flurry of action and sound. The hunter in the covert moved up behind the dogs. With a metallic "whrrr" a long-billed timberdoodle broke for the edge. The covering gunner swung ahead of the twisting target and missed. His second barrel tumbled the woodcock. It dropped, leaving a puff of russet feathers drifting in the air. At the command "fetch," the setter romped forward and brought the bird to his master, who pocketed it after letting the eager pup mouth the kill.

The same drama, with variations, repeated itself several times before they reached the end of the covert. Once each dog pointed different birds, one of which got cleanly away. At another point an old cock grouse, wise in the ways of hunters, scuttled unseen and unheard ahead of the dogs, laying a trail of scent. One gunner, circling the edge of the cover, intercepted the wily old-timer as he thundered for the safety of a spruce thicket beyond the point of the covert.

With dogs at heel and game pockets thickened by four woodcock and a grouse, the gunners trudged down to their car. In the fading light two woodcock whistled by, silhouetted briefly against the evening sky as they fluttered into the alders. The sound of other more distant birds tumbling into the hardhack of the fields gave promise of future sport to come.

On the same day, other hunters and other dogs—in alder and aspen

Figure 36. Hunting cover, Michigan (Photo by Michigan Conservation Dept.)

coverts from Minnesota to Nova Scotia and in the crabapple and hawthorn thickets of Pennsylvania—had tasted the finest upland hunting that the North can give.

Woodcock hunting on the principal breeding grounds usually goes hand in hand with Ruffed Grouse hunting. Both species often frequent the same coverts. Some are primarily woodcock coverts, and many fine grouse coverts have trees too big for woodcock. Many grouse hunters take woodcock incidental to their favorite sport. A small minority of hunters, however, seek woodcock by preference, and consider hunting it the finest upland sport (Figure 36). The woodcock lies well for a dog and exudes strong scent. When flushed, it presents a fast and tricky target.

Equipment

Any shotgun with a relatively open bore and from .410 to 12 gauge can be used for woodcock hunting. The smaller the shot size the better. Number 9's are preferred by most hunters and they are heavy enough to use on grouse early in the fall.

Dogs

The fascination of hunting woodcock is lost if it is not a man-dog team effort. Dogs must be used to find woodcock consistently. The pleasure of watching a good dog hunt, the anticipation as it "makes game" or points, provides the essential thrill and beauty of this sport.

When it comes to a choice of dog breeds, most woodcock hunters have definite preferences. There are good and poor woodcock dogs in every breed. Almost all sporting dogs have been used to hunt woodcock. Some favor the pointing breeds, such as the English pointer and setter, Brittany spaniel, Gordon setter, field-bred Irish setter, German shorthair pointer, and Weimaraner. Others swear by the close-working flushing dogs—American and English Springer spaniel or one of the breeds of retrievers.

Figure 37. Author's dog pointing brood

I prefer the pointing breeds because there is more anticipation of a shot over a staunch dog. Some hunters like a close-working, cautious dog, while others favor a faster, wider-ranging dog which they usually hunt with a light bell. Speed must be coupled with a good nose and staunchness. One of the best dogs that I used in this study was an English setter bitch (Figure 37). She learned to hunt woodcock before she was a year old. She was wide-ranging and fast, but she was extremely dependable. A woodcock dog must be well trained and handled with judgment and understanding. Some require stricter training than others. Methods of training may be basically the same, but the technique of enforcing orders varies with the temperament of the individual dog.

I look for dogs with a highly developed pointing instinct, with brains and tractability. For success, hunting experience is as important for the dog as it is for the hunter. A few dogs, capable on other game birds, will not point woodcock. Many dogs find dead woodcock unpalatable and refuse to retrieve them. In breaking a young grouse dog the woodcock adds welcome variety, since many young dogs can handle it. The woodcock is a steadying influence on an inexperienced dog which may be rattled by the running, quick-flushing grouse.

Hunting Success

Of all states and Canadian provinces, the woodcock comes nearest to being the preferred upland bird in New Brunswick and Nova Scotia. To the best of my knowledge, New Brunswick is the only place where field trials are run on woodcock.

Hunting success depends, of course, on the gunner's eye, the ability of the dog, the type of cover, and on the local abundance of birds. Studies in New Brunswick (Wright, personal communication) and Pennsylvania (Liscinsky, 1962) report varied success. Often between 40 and 50 per cent of all birds flushed are killed. I have seen good shots kill 9 out of every 10 birds flushed. Under other circumstances, the kill was as low as 20 per cent. If resident birds are present, early in the season, a team of expert hunters with a good dog can virtually clean out a small covert in only a few visits. Wright (letter) said that one "party" may take 35 per cent of birds in a covert. He believes that out of three birds present, one is killed, one missed, one not seen. He estimated crippling loss at 1.5 per cent; the percentage of birds flushed which were killed varied from 0 to 78, based on 26 hunts in 9 coverts. Liscinsky (1956) found from 1952 to 1956 that hunters in his study areas killed between 43 and 50 per cent of the birds flushed. Wright also studied dog efficiency, discovering that the average dog points slightly more than half the birds flushed.

The most precise information of hunting success has been supplied me by G. A. Ammann (letter) of Michigan. He cites the record of a Michigan hunter, Edward A. Smith, who hunts woodcock and grouse. Smith's shooting average was 24 per cent of 1140 woodcock flushed 1959–1964 (Table 3). He flushed an average of 2.8 per hour.

TABLE 3. Woodcock Flushed and Shot in Michigan by Edward A. Smith

	1959	*1960*	*1961*	*1962*	*1963*	*1964*	*6 yr. Av.*
Hours hunted	78	76	75	67½	70½	65½	72
Woodcock flushed	191	167	190	205	180	207	190
Woodcock shot	40	52	42	39	55	51	46

Smith's flushing rates were considerably higher when he and others were hunting woodcock primarily. Ammann supplied the following records of Michigan woodcock hunts (Table 4). Blankenship compiled reports of Michigan cooperators who only hunted woodcock in 1954 and 1955. The results are based on 1171 hours of

TABLE 4. Michigan Woodcock Flushes

	Hours Hunted	*Woodcock Flushed*	*Woodcock Flushed per Hour*
Smith (1960)	22	115	5.23
Goudy (1964)	24½	141	5.82
Prawdzik (1964)	39¼	230	5.85
Ammann (1964)	24	208	8.67

hunting and 1975 woodcock flushed. Five hundred and fifty three of the latter were killed, giving a hunting success of 27 per cent. Approximately one of every four woodcock flushed in Michigan coverts is killed.

Importance as a Game Bird

Each year, American hunters shoot at least as many woodcock as they do some species of waterfowl. The woodcock is not a bird for the meat hunter; the sport and recreation offered are more important than the dinner menu. Besides being shot for sport, the bird has unusual aesthetic interest for bird watchers and ornithologists.

Estimates of kills are given in Table 5. Kill records from Michigan

TABLE 5. Estimated Woodcock Kill Based on Most Recent Reports (1969)

State or Province	*Estimated Kill*
Alabama	Unknown—minor
Arkansas	Unknown—minor
Connecticut	22,630
Delaware	Unknown— 1,235 reported in 1935
Florida	1,000
Georgia	250
Illinois	14,000
Indiana	8,155
Kentucky	42,000
Louisiana	10,000
Maine	91,870
Maryland	17,156
Massachusetts	32,230
Michigan	160,840
Minnesota	594 (sample only) 3,000 *
Mississippi	Unknown
Missouri	5,000
New Brunswick	6,000
New Hampshire	10,000 *
New Jersey	60,000
New York	138,000
North Carolina	8,401
Nova Scotia	17,000 (1957)
Ohio	3,500
Oklahoma	Unknown—minor
Ontario	5,000 *
Pennsylvania	65,000
Prince Edward Island	500 (1958)
Quebec	7,400
Rhode Island	374
South Carolina	1,000
Tennessee	500
Texas	1,000
Vermont	10,000 *
Virginia	Unknown—minor
West Virginia	500 *
Wisconsin	47,800
Total	788,935

* Although no estimates are available, most of these areas are heavily hunted for woodcock, and the figures are considered conservative.

and New York appear more reliable than estimates from other states. Several states base their estimates on game kill card reports, and a few use hunter questionnaires. The validity of many is questionable, but they represent a minimum. Woodcock are killed incidentally to quail and rabbit in several southern states, but woodcock hunting is growing more popular in the South.

The list records 788,935 kills. We can assume that hunters kill more than half a million birds each year.

Fall Habitat

The successful hunter soon learns to recognize woodcock coverts, good habitat having qualities that an experienced gunner can spot at a glance. The requirements of woodcock are more stringent than those of most game birds. But the pattern of vegetation that makes woodcock habitat varies greatly in different areas of the range. To use its long prehensile beak the woodcock needs moist soil rich in earthworms. It requires openings in the forest particularly during the breeding season and more or less throughout the year. To thrive the woodcock must have young forests of proper density, containing woody plants a few inches high to trees little taller than 30 feet. The density of vegetation six inches above the ground must be relatively light for easy foraging. In hot weather woodcock like to rest in dense shade, particularly under evergreens. At the height of the migration they may stop almost anywhere, but their living areas include the features described above. Many coverts appear to meet woodcock needs, but close examination often reveals unsatisfactory conditions. Young aspens or birch usually indicate good woodcock cover, but these trees often grow on thin soil where the forest floor is covered with moss, offering no attraction to woodcock. Fall coverts differ little in their essential vegetative components from those of early spring.

Clarke (letter) found that woodcock preferences for soil types were not important in spring and early summer in Ontario, but were important in autumn. At that time of year Ontario woodcock are concentrated on two types. The first is porous and rich gleisol, and the second dolomite or limestone, where the soil is too shallow for the plow. In the latter soil type the cover is usually pastured and there are depressions trapping moisture where organic desposits accumulate.

Mendall and Aldous subdivide fall cover into "early" and "late" classes. From September 15 to October 15 in Maine 45 per cent of all flushes recorded were in alder, 35 per cent in second-growth hardwoods, 17 per cent in second-growth mixed stands, and 3 per cent in brushland. As the fall advanced, alder became more important, accounting for 63 per cent of the flushes. Twenty-nine per cent of the flushes occurred in mixed growth and 8 per cent in hardwoods. Bennett and English grouped the central Pennsylvania fall coverts into two types—alder bottomland and wild crabapple and hawthorn thickets on mountain slopes.

The vegetation on fall flushing sites has been studied in detail by Liscinsky (1964) in central Pennsylvania and Wright (letter) in New Brunswick. In most of the Northeast, immature alder and aspen are the most common trees in woodcock coverts. In his analysis of 28 aspen stands in Pennsylvania, Liscinsky states that, in per cent of density, aspen rated 48, followed by scrub oak with 16 and hazelnut with 10. Other components of the Pennsylvania aspen stands in order of importance were hawthorn, black cherry, blueberry, dewberry, and red maple. Pennsylvania alder coverts also contained, in order of frequency of occurrence, silky dogwood, blackberry, and black willow. Liscinsky lists 24 other woody plants commonly associated with alder.

Wright studied 19 productive covers in New Brunswick. He found the most important tree species to be alder and gray birch, followed by evergreen species. Alder often invades the denuded areas of old beaver ponds. The principal herbaceous plants included grasses, bracken, raspberry, and goldenrod. One cover contained beech as the dominant species. Blankenship found that summer and fall covers were similar, except that woodcock prefer lighter cover in the autumn. In frequency of occurrence, the three most important trees were alder, aspen, and spruce.

In Massachusetts the picture is similar to that in other northeastern states. The principal flushing sites are dominated by alder, followed by aspen and gray birch. Scattered evergreens in the understory seem to raise the carrying capacity of all Massachusetts coverts, and the same is true of approximately 50 areas in Vermont that I have hunted for 25 years.

The best woodcock habitat has trees from 10 to 20 feet tall. Although alder and aspen are the most important species in the Northeast, there are exceptions. Martha's Vineyard Island, off the coast of Massachusetts, has many resident woodcock. The dominant woody vegetation in its woodcock coverts is black locust, followed by tree of heaven, another fast-growing species, and greenbrier. Nantucket Island woodcock coverts are dominated by highbush blueberry.

During the fall woodcock may rest on well-drained sites during the day. I have often found them on side hills with small white birch, meadowsweet, or hardhack edges. Even the so-called "aspen" or "alder" coverts contain a variety of other plants. Pure alder or aspen stands are not so attractive to woodcock as coverts where alder and aspen are mixed. The most productive covert I ever hunted grew on a hill in Maine and was formed by ideal juxtaposition of young alder, apple, aspen, gray birch, and white birch, sprinkled with young white pines and broken by small openings. Liscinsky (1964) points out that the composition and distribution of cover in Pennsylvania are probably more important than the cover type. Large even-aged stands of one species are less useful than small stands of mixed alder, aspen, crabapple, hawthorn, gray dogwood or silky dogwood.

Ammann (personal communication) has always found large numbers of woodcock in early fall on Drummond Island on northern Michigan. He describes coverts in this area as "a considerable mixture of conifers (all sizes), aspen, and patches of heavily browsed red-osier dogwood." He stresses the importance of conifers on this area, especially young stands of spruce and balsam. In most of Michigan the majority of woodcock found in the fall is in "alder swales or mixtures which include alder."

Signs

The hunter can often determine if woodcock are present by two signs. The most conspicuous are droppings referred to by hunters as "chalk." These are about the size of a 25-cent piece, white with brown streaks. The hunter will also often find a series of probe holes where the woodcock has been "boring" for worms.

Age of Habitat

Woodcock cover is transitory, young forests becoming too old to attract woodcock. Most coverts pass from excellent habitat to overgrown woodland in 25 years. Many hunters continue to revisit bygone productive coverts. They find fewer and fewer birds as the years pass, often regarding this as evidence of a population decline.

They fail to recognize habitat change has prompted the birds to move elsewhere.

I was able to document the history of a covert in Bethel, Vermont, for 25 years. It was abandoned as a hayfield in 1935. Within ten years forest crept into the 50-acre field in the form of white birch, aspen and white pine. In 1946 and 1947 the area contained excellent woodcock habitat. As the years passed and the wooded area expanded, the birds occupied the fringes of the woods where the trees were between 10 and 20 feet high. By 1960 few woodcock used the 1946 cover. Fingers of woody vegetation extending over the old hayfield had become the favored habitat. Eventually the field will be smothered with large trees, and the woodcock will seek more favorable living areas.

Weeden (1955) studied woodcock use of alder coverts in relation to tree age in Maine. The average age of unused stands was 11.1 years, and of used stands, 7.7 years. Alders grow most rapidly in their first 7 years and live only about 20 years, though most alder stands are of uneven age. Wright reports alder coverts that have yielded good woodcock hunting for 25 years, and I have found woodcock in some alder coverts in central Vermont through 20 years. On two areas, one in central Vermont and another in central Massachusetts, sections of alder coverts that yielded woodcock 10 to 20 years ago are no longer used, but the alder growth during this time has invaded surrounding abandoned fields where the age and height are suitable for woodcock. In short, if abandoned land suitable for alder seeding adjoins an alder covert, young attractive alder patches will gradually replace the older sections as woodcock habitat. If a covert is surrounded by woods so that new stands of alders cannot take root, woodcock will gradually abandon it after the eighth year of growth.

Cover Density

Blankenship laid out 25 random plots in each of 13 woodcock coverts and analyzed the vegetation according to species and density. He separated the plants into three broad groups as follows: Trees—any woody plant over 10 feet high; Shrubs—woody plants between 1 foot and 10 feet high; Ground Cover—woody plants and herbs 1 foot or less in height. Average cover densities of these vegetation groups for the nine feeding areas and the four loafing areas are recorded in Table R, Appendix II. Blankenship admits that the apparent density of ground cover is misleading, since there were enough bare spots or low density plants to allow freedom of movement by foraging woodcock.

Liscinsky described ideal woodcock cover in Pennsylvania as having a density of 75 per cent in overstory and 25 per cent in ground cover. I have flushed hundreds of woodcock from coverts with all degrees of overstory density. Loafing areas in conifer plantations sometimes have 100 per cent tree crown closure, and a ground density as low as 50 per cent. In stands of hardhack or blackberry or green brier density three to four feet above the ground often approaches 90 per cent, but the ground density may be only 50 to 60 per cent. Bracken fern is a common herbaceous plant is some woodcock coverts; the fronds, one foot above the ground, shade 70 per cent or more of the forest floor.

If coverts are not grazed, the alder stands must be dense enough to shade out herbaceous growth. Grazing by a few cattle will keep ground cover sufficiently sparse to attract woodcock, even though the alder density may be as light as 50 per cent.

Other Habitat

I have often flushed woodcock under apple trees in abandoned orchards where the soil is enriched by several generations of rotting apples. Knight (1944) found apple trees to be an important component of Pennsylvania woodcock cover. In 1964 during a drought in central Vermont, I found many woodcock in fertile abandoned fields grown up to white pine about 20 feet high. Hardhack was scattered in the edges and in the few openings of the pine stand.

In the fall of 1964, a graduate student flushed 12 woodcock in mid-October from an uncut cornfield with stalks 4 feet high. This field was surrounded by alders and other good habitat. Most cornfields are rich in earthworms and probably are frequently used in migration. In late September, 1964, Thompson (1966) flushed eight woodcock in mid-afternoon from the mud in a marsh of chest-high bulrush. The birds were between 30 and 70 yards from the nearest surrounding cover. In coverts of all types woodcock are more apt to be found near open edges than in the middle of the covert.

CHAPTER IX

Management

The inconspicuous woodcock is a neglected stepchild of wildlife management. Most state game administrators consider it a game bird of minor importance. Yet, in terms of numbers bagged and hunter hours of recreation, it is as important, within its range, as the Canada goose (*Branta canadensis*). Woodcock hunting ranks high in quality and low in the quantity of the meat.

Woodcock hunting can become almost an addiction. Once fully tasted, it gets into the blood, and no amount of hunting of other kinds can ever quite purge it. It offers recreation as unique as the bird itself. Those who have pursued the russet-colored timberdoodle in the October uplands regard the sport as the most rewarding available. Few of us who know this thrill feel that the funds now spent by administrative agencies to manage the woodcock begin to match its real and potential values.

The chief management tool has been that of setting seasons and bag limits based on the spring census of courting males as an index. The validity of this method is open to question without greater standardization, and other measures of population fluctuations are constantly being tested.

CENSUS

Spring Census

For the last 25 years biologists have established an index of annual woodcock populations by counting the number of singing male birds heard over a prescribed route in woodcock habitat. The pertinent instructions sent to cooperators by the Federal Bureau of Sport Fisheries and Wildlife are:

1. Select route in area known to be woodcock habitat.

2. Begin census after migrants have passed through and complete it before peak of hatch.
3. Start count at stop where there is known to be a singing bird.
4. Proceed on route with stops no closer than .4 of a mile.
5. Stop for two minutes at each listening post.
6. Count peenting males and birds in courtship flight which can be distinguished from peenting males.
7. Duration of run should be 35 minutes.
8. Avoid runs during following weather conditions:
   a. Two days before, during, and one day after full moon.
   b. Temperature below approximately 40 degrees.
   c. Winds eight or more miles per hour.
   d. Heavy rain or snow.

Variables affecting the census techniques are many (Chapter V). No one bird sings the same length of time each night, even in the same weather conditions. The greatest variable is the range in hearing of different people. Tests in Massachusetts showed that some observers could hear twice as many birds, or more, over the same route during the same evening as others.

The most serious error in this method lies in not recognizing the relatively rapid change in vegetation, which results in the habitat being unsuitable for woodcock. Constant vigilance of cover changes on census routes should be required. In one route in Quabbin Reservation, Massachusetts, counts decreased from 67 to 16 over six years. In my opinion this did not reflect a population decline, but a changed habitat. Recognition of such changes requires extensive field experience. During the same period another route ten miles away, where vegetation had been kept cleared, showed a slight increase in counts of occupied singing grounds.

Kozicky *et al.* (1954) undertook a statistical analysis of the singing ground count method. They analyzed forty-four routes in northeastern United States. Their resulting recommendations included changing the sampling unit from mean number of occupied singing grounds per route to the mean number of singing birds per stop, which is the method now employed. In spite of variations in counts over the same route, they demonstrated the sampling error could be reduced by devoting the same time and energy to increasing the number of routes and eliminating multiple runs per route. Kozicky's

Figure 38. Flushing woodcock (Photo by Richard Borden)

recommendation that increased numbers of routes should erase many variables, appears sound. The validity of the annual index from any region is directly proportional to the number of routes run and the experience of the census taker.

Census routes were not selected on a random basis, regardless of woodcock habitat, until the spring of 1962 when Martin, with the aid of cooperators, selected and censused 126 routes in Michigan on an experimental basis (Martin, 1963 and 1964). Counts on the random routes in 1964 showed a slight increase in birds in northern Michigan and a decrease in southern Michigan, whereas the numbers tabulated on the operational routes showed an opposite trend. Random routes must be run several more seasons to assess their validity (Martin, 1964). The adoption of Kozicky's recommendations, and even the use of random sampling, requires manpower. In Massachusetts it is rare to have favorable weather in more than half the evenings during three weeks when routes must be run (following migration and preceding decrease of breeding activity). One man should not be assigned more than ten routes a spring, even though he could run more in a spring of unusually favorable weather.

I believe that any drastic decline in woodcock populations such as occurred in the spring of 1940 following a bad winter freeze, would be reflected in the annual spring censuses although a smaller but still significant decline might well go undetected.

Other Methods

Ammann (letter) computed woodcock flushed per hour from a random poll of Michigan woodcock hunters from 1954 to 1963. The count of each previous spring was also listed. There is no correlation between the spring count and the birds flushed per hour between 1954 and 1960, with the possible exception of 1959. From 1961 to 1963, there appears to be some correlation in the spring and fall figures (Table 6).

TABLE 6. Woodcock Flushed Per Gun Hour Compared With Spring Counts and Computed Kills in Michigan

	Woodcock Flushed per Gun Hour	*No. of Hours*	*Computed Woodcock Kill*	*Spring Counts: Mean Birds per Stop*
1954	.35	3598	49,320	1.00
1955	.51	3069½	46,230	.90
1956	.49	2524	40,630	.83
1957	.55	2562	67,760	.93
1958	.49	5176	62,430	.98
1959	.58	4941½	64,460	1.16
1960	.54	5528	83,890	1.02
1961	.62	7562¼	102,920	1.10
1962	.71	7951¾	118,580	1.18
1963	.69	88[illegible]3	144,940	1.33
1964				1.34

Reid and Goodrum (1957) conducted woodcock censuses with bird dogs in the piney woods of west-central Louisiana annually since the winter of 1949–50. Their results demonstrated that the use of a dog is a possible method of counting wintering woodcock in the pinelands of Louisiana. Since weather in the South seriously affects distribution of woodcock in the uplands and lowlands of wintering areas such censuses may not be reliable.

Figure 39. Alders in lumbered fir and cedar swamp (Photo by Eldon Clark)

On January 22 and 23, 1952, Reid (Reid and Goodrum, 1953) spent 9½ hours working dogs in shrubby growth and swampland adjacent to nighttime feeding fields at Swayzee Lake in St. Landry Parish. No woodcock were located by Reid on those dates; yet, on January 27, 156 woodcock were banded and many others flushed by Louisiana State University personnel in the Swayzee Lake fields (Glasgow 1958). The sandy soils in these fields, which had dried out prior to Reid's census, were moist from heavy rains on January 26 and 27. Throughout this study it was noted that a lack of precipitation for a period of two weeks resulted in a drastic decline in the number of woodcock using nocturnal feeding fields. Conversely, on numerous occasions above-average numbers of woodcock were found in adjacent areas having sufficient moisture.

Glasgow (1958) made a number of attempts to establish an index of population trends by counting woodcock on wintering fields. He found that dragging long ropes through the fields at night or placing men at strategic places at the perimeter of the fields to count woodcock both proved impractical. He thought that perhaps woodcock flushed on a man-hour basis held some promise as a technique that might give a reliable indication of woodcock numbers. He tabulated woodcock on this basis, but pointed out inconsistencies from one year to the next traceable to vagaries of the weather. For example, in 1950–51 moisture was deficient in December. A heavy freeze occurred in late January forcing woodcock to abandon feeding fields. This caused a decline in woodcock flushed on a man-hour basis. In 1951–52, with favorable weather conditions, there was a gain in flush rates of 88 per cent. In later years Glasgow refined the technique further. He weighed weather factors on the basis of his long experience and felt that woodcock flushed per man-hour had some validity as a population index. He experimented once in counting woodcock at night from horseback, and found, if the objective was counting and not banding, that this was the most effective method of direct count. If experienced manpower were available so that simultaneous counts could be made on representative woodcock wintering habitat in both uplands and lowlands, a reliable winter index could be established.

Creating and Maintaining Habitat

Woodcock management has been tried experimentally by some other workers. The usual practices include arresting vegetative succession, cutting or burning forest tracts, grazing, herbicide treatments and planting. The most intensive effort has been at the Moosehorn National Wildlife Refuge in Calais, Maine. Moosehorn is the only refuge which was established primarily for woodcock. Strip cuts of 500 blocks containing 246 acres, and experimental clearings were made in 1940 and 1941. Sufficient time has elapsed since then to assess the effectiveness of those methods in maintaining woodcock habitat.

The Moosehorn operations have been handicapped by three factors. In the first place, there had been no previous experiment to serve as a guide in creating woodcock habitat. Second, a large deer herd continually browsed back sprout growth where cuttings were made. And third, even today no statistically sound method has been found to measure woodcock use or population changes accurately on the improved areas.

In 1940 the refuge managers created 37 artificial singing grounds on the Maine tract, ranging in size from .15 to 1.94 acres. Many were inadvertently maintained by browsing deer, and all were subsequently used by courting male birds.

Since 1940 approximately 800 acres on the refuge have been cut to create woodcock coverts. Clear cutting, thinning, burning, bulldozing, flooding and grazing have been tried. Thinning and bulldozing proved most effective in shrub areas or in meadows with heavy grass. Clear cutting was the most successful method in older forest stands. Cutting on dry upland covers with relatively thin soil has proved unsuccessful in creating woodcock habitat. Clear cuttings of balsam and cedar on low moist areas, however, were invaded naturally by new alder coverts (Figure 39).

Reardon (1950) made a study of woodcock use of several cover types which had been cut on the refuge in an effort to attract woodcock. The types were classified as "pure" alder, mixed growth with alder predominating, pure hardwood, and mixed stands of hardwood and softwood. The criteria for measuring woodcock abundance were the number found by bird dogs and the amount of woodcock sign observed. Male birds utilized the openings of the developed coverts for singing grounds. Woodcock used the managed areas much more extensively in summer and fall than they did the nearby unimproved cover. Reardon found that clearings on valley coverts contained more woodcock than similar clearings in the uplands. This may well have been traceable to differences in moisture on the two kinds of sites rather than to type of cuttings.

Liscinsky (1962) has worked extensively in Pennsylvania to create or maintain woodcock habitat. He has tried chemical spraying, cutting, pasturing, cultivating, and planting. Not enough time has elapsed to accurately evaluate several of his methods. On his pastured areas, steers trampled underbrush and created clear trails which were used by woodcock.

Blankenship tried improving woodcock feeding sites in Michigan by chemical spraying. In three areas where heavy ground vegetation precluded much use by woodcock, he laid out 10 plots one-hundredth of an acre in size. Dalapon was sprayed with a hand-pump garden sprayer at the rate of 35–55 pounds per 50 gallons of water per acre in July 1955. A year later the plots were barren of grasses and sedges while surrounding areas approached maximum density. By using statistical methods ("t" test) to evaluate counts of birds and sign,

Blankenship found that woodcock used the treated areas much more intensively than they did the control areas.

Controlled burning may be effective, depending on the soils, vegetation and seed source. In some of the Lake States and elsewhere, aspen is the first tree to grow up in burned-over land; in Massachusetts, gray birch usually succeeds fires. From a practical point of view, of course, burning forests with potential commercial value could never be recommended. There are, however, many wooded areas on thin soil which have little commercial potential. Burning such areas might be justified to create woodcock (and Ruffed Grouse) habitat. Burning in any tract should be applied with caution, and preceded by burning small experimental plots.

In one forest tract belonging to the University of Massachusetts, various methods were used to kill standing hardwoods of low value. Basal poisonings, aerial spraying, and cutting rings were all tested. Ground spraying with 2–4–5D for trees up to 30 feet high (MacConnell and Bond, 1961) appeared to be the most economical method. Woodcock used openings around the sprout growth resulting from these practices. Clearings of this type should preferably be near a stream or in damp sites.

In Quabbin Reservation, Massachusetts, the view from one section of road above the reservoir has been enhanced by cutting back bordering vegetation so the reservoir will not be masked. The number of singing males in this area has not decreased for twelve years. In one area, a clearing created a singing ground in a tract that formerly contained a dense stand of hardwoods forty feet high.

Land clearing, however, can be carried so far that it reaches a point of diminishing returns from the woodcock's viewpoint. A 20-acre field which formerly supported dense stands of young alder and highbush blueberries was mowed and seeded to grass to create picnic areas; only three courting fields remained where formerly there had been eight to ten.

Liscinsky (1965) in Pennsylvania found that it is relatively easy to plant alder in moist sites. Alder coverts can be established by planting seedlings and by direct sowing of seed. Two- and three-year-old seedling stock has been found to be the best for planting. Greatest success in direct seeding was achieved when fall-collected seed was sown the following February or March. Discing of the

Figure 40. Alder cover, Massachusetts

sites prior to planting produced better results, but catches were good even when the seed was sown directly.

Aspen can be planted easily, but must have full sunlight to reproduce and grow in abundance. Clear cutting of over-mature aspen results in fine sprout growth. Discing the soil prior to cutting accelerates sprout growth.

Most other trees or shrubs attractive to woodcock can be planted. Gray dogwood, for example, has the ability to grow on a wide variety of sites and produce abundant fruit of value to other species of wildlife. Fair success has been realized from planting gray dogwood seedling stock and direct seeding. This species requires full daylight and fertile soils.

Scattered stands of conifers—especially spruce or pine in scattered clumps or singly—are used by woodcock as nesting and loafing areas. Most species are easily planted from seed or two-year-old nursery stock.

Alder coverts, and young aspen coverts, where crown closure is tight enough to prevent the growth of dense ground cover, are attractive to woodcock (Figures 40, 41). As the trees mature and some die, light reaches the forest floor, and heavy grass or other vegetation invades the site. Unless there are bordering abandoned fields where seedlings can grow, the covert receives little woodcock use. Cutting or poisoning is then required to restore good woodcock cover in the area. Where alders are more scattered, of uneven age, and have an undergrowth of heavy grass, pasturing of cattle would seem to be the most practical and economical method of management.

In a ten-acre study area near Amherst, Massachusetts, records of woodcock usage were kept for twelve years. In the first four years the area was grazed, and eight to twelve birds were flushed in fall or spring. At least two broods were reared there each year. After four years the livestock were removed and some of the alders cut. Usage has steadily declined. In 1961, during several visits, I found only two woodcock.

Mendall and Aldous (1943) suggested creating refuges of several thousand acres along the woodcock migration routes. Quabbin Reservation in central Massachusetts is closed to hunting and thus comprises a refuge of 88,000 acres. Traditional stopping grounds, such as Cape May, New Jersey, where woodcock are concentrated and vulnerable, are logical sites for refuges. The few acres protected on Cape May by the National Audubon Society are probably more effective than several larger refuges in other areas.

HUNTING SEASON AND BAG LIMITS

The only mortality factor that can be controlled by man is the gun. Because the woodcock is a migratory bird, hunting seasons and bag limits are set by the Federal Government. In times of catastrophic winter mortality the government can immediately curtail the season. In 1940, following severe freezes and woodcock deaths on the wintering grounds, the season was reduced from 30 days to 15 days. Woodcock populations subsequently recovered, although no one could measure whether this was a result of shortening the season.

Although the government establishes the length of the season, it is left to the option of the states to select this period within the broad limits of September to January. In New Brunswick the season opens on September 15; in the northern New England states and the Lake states woodcock can be hunted on October 1, October 10 in Massa-

Figure 41. Ideal aspen covert

chusetts, October 20 in New Jersey, and at later dates in more southern states.

I believe that the opening dates set by the states may have more effect on woodcock populations than the length of the season. When my research revealed that many migratory woodcock passed through Massachusetts before the opening of the hunting season on October 20 and were gone two weeks before the closing date of November 29, the season was advanced to an opening date of October 10. The calculated kill of woodcock in the state then increased from approximately 12,000 to 30,000.

It is possible that local native populations may be seriously depleted by hunting if subjected to heavy gun pressure. Liscinsky (1964) cites documentary evidence that shooting seriously reduced a native woodcock population in Pennsylvania when the opening date of the hunting season was October 1. An area in Luzerne County had a history of growing and persistent hunting pressure reflected in a decline of the woodcock population. In 1956 the study tract was closed to hunting. From 1956 to 1960 when there was no shooting, the estimated fall numbers increased from 18 to 60 woodcock. Since

1961 this area was reopened to hunting and the opening date of the season was changed to mid-October. Liscinsky states: "To date [1964] the number of woodcock bagged has remained stable at a rather high level" (about 30 birds). Martin (1963) evaluated the effect of regulations on size of kill from the 1959–61 woodcock wing surveys. From the states where sufficient data were gathered, only seven per cent of the total kill as indicated on envelopes sent in by hunters contained the daily bag limit of four wings or more. Many envelopes contained wings from parties of hunters, so interpretation of kill on this basis may be misleading. The tentative results suggest that relatively few hunters bag four birds a day. Martin estimated that an increase in the daily limit to six birds would increase the total kill by no more than ten per cent. In 1963 the bag limit was increased from four to five birds. Ammann (personal communication) estimated the liberalized regulations resulted in a kill increase of two to three per cent in Michigan. Woodcock appear to be holding their own with a current season of 50 days and the increased bag limit.

EFFECTS OF LAND CHANGES

Certain land practices have provided us a key to those which attract woodcock and those deterring them. Heavy logging and light forest pasturing near open cultivated land may create woodcock habitat. Evergreen plantations are attractive to breeding birds during the first five or six years of growth. The recent spread of the beaver has favored the woodcock, because beaver colonies usually develop openings that are invaded by alders and other choice woodcock cover.

In the alluvial bottomlands on the woodcock's winter range in Louisiana, much drainage of wet land has been subsidized by the Federal Government. Some federal agencies provide the landowner with free technical aid and charge him only a nominal fee for carrying out complete drainage of his land. This has destroyed many acres of woodcock habitat. Glasgow (1958) believes that drainage may continue until all swamps that can be drained are drained. When this point is reached, swamps that lie too low to drain will probably be diked and the excess water pumped off.

Drainage destroys both daytime cover and night feeding sites for woodcock. When a low swampland is drained, the water table is lowered on adjacent, higher agricultural land. Consequently, many of the low wet pastures along the edge of the swamps, when better drained, are converted to crop land. Land that remained moist before drainage now dries in a few days. Since the alluvial soil dries extremely hard, it is then impossible for woodcock to probe in it.

Some southern forestry practices improve woodcock range while others are harmful (Glasgow, 1958). Burning of pinelands to remove an accumulation of "rough" is likely to improve the habitat unless the fire destroys too much of the brush. Elimination of hardwood shrubs and brush should be discouraged. Converting mixed hardwood-pine stands to pure pine reduces the value of the forest as a woodcock habitat. Young hardwood thickets provide excellent daytime cover. Forested stream bottoms should be left in their natural condition. Since forest production is the best use for much of the hill land in Louisiana, woodcock are expected to have sufficient wintering range during mild winters despite the drainage of the lowlands.

Increasingly efficient farming operations have reduced the number of brushy fence rows, briar patches in pastures, and brush-covered drains where formerly woodcock would always be found during the winter. Pasture improvement practices have eliminated much of the weedy growth on night feeding sites, and fall plowing and the planting of winter cover crops have reduced the value of land for feeding sites. Mechanized equipment permits the operator to cultivate more acres, and so few fields remain in a fallow condition. These efficient farming practices are generally detrimental to woodcock.

CHAPTER X

Population Dynamics

If one assumes the hunter kill of approximately 800,000 represents 10 per cent of the continental population, the total population would be approximately 8,000,000 birds. This is an assumption with no specific supporting data.

The American Woodcock hen can lay only one clutch of four eggs a year. Thus, the woodcock will survive as a species only so long as its annual mortality remains low. Fortunately, although its rate of replacement of annual losses is limited (approximately 50 per cent), the woodcock is blessed with an average life span longer than those of most other upland game birds. If woodcock reach adult size, life expectancy is 1.8 years (Sheldon, 1956). Eighty per cent of all fall Ring-necked Pheasant may be birds of the year (Edminster, 1954). From 75 to 80 per cent of all autumn Bobwhite Quail (*Colinus virginianus*) were hatched the previous spring (Ripley, 1957). Quail and pheasants, however, are compensated by a larger clutch size; they lay three times as many eggs as woodcock.

Our knowledge of woodcock reproduction and survival is sketchy. But the door to this vital subject has been unlocked during the last few years, and I believe we now have reliable ways to measure annual productivity and to learn whether the woodcock is threatened. If there is a drop in the number of juveniles in the fall, it is reasonable to assume fewer birds will be returning north to breed the following spring.

I developed original techniques in Massachusetts which provided the keys to further understanding of population dynamics. The spring trapping technique provided valuable data on annual survival of males. Summer netting yielded information on molting which in turn led to a study of differences in the wings of juveniles and adults in the fall. We first detected that the ages could be separated by the wear on the outer three primaries (Sheldon, Greeley, and Kupa,

1958). As a result, the Bureau of Sport Fisheries and Wildlife solicited thousands of wings from woodcock hunters. Martin (1964 and Appendix) further refined and accelerated the aging process by recognizing a color pattern difference in the secondary wing feathers. This discovery made it possible to recognize yearlings before their molt in the subsequent spring and summer. Sex can be determined by the width of the outer primaries (Greeley, 1953). These methods have resulted in the most reliable and extensive measure of woodcock productivity available.

Hunters between 1959 and 1962 sent the Bureau of Sport Fisheries and Wildlife more than 40,000 woodcock wings. I have analyzed the sex and age of 34,617 wings from states with the largest number of specimens—Maine, Michigan, Wisconsin, New York, New Hampshire, New Jersey, and Pennsylvania—and the Canadian province of New Brunswick (Table S, Appendix II). Martin (1965) recently published the results of analyses of 45,444 wings which were not significantly different from the sample I had available.

SEX RATIO

The continental sex ratio was 82 males : 100 females. When the sample is broken into age groups a high contrast in sex ratios appears. In adults there were 6,402 males and 10,170 females, a ratio of 63 males : 100 females. Among 18,045 juveniles the sex ratio was virtually even: 9,170 males and 8,875 females (103 males : 100 females). Sex ratio computed separately for each of the four years did not differ from the combined totals.

More limited collections by other workers also contained a preponderance of females. Mendall and Aldous found a ratio of 74 males : 100 females among 669 birds from Maine, New Brunswick, and Nova Scotia. Greeley analyzed the sex of 209 Wisconsin birds which contained 80 males : 100 females; Blankenship sexed 207 Michigan birds and found 86 males : 100 females. The largest sample, collected by Glasgow in Louisiana, contained 5,169 birds with 70 males to each 100 females. This sample may have been somewhat biased since more females were captured in late winter when some of the males had preceded them on migration to the breeding grounds. None of the earlier analyses given above were separated into age classes (Table 7).

The unbalanced fall sex ratio of adult woodcock with a preponderance of females is the opposite of that of many upland gallinaceous birds in which the fall populations usually have more males than females (Table 8).

TABLE 7. Sex Ratios of Woodcock from Various Parts of the Range

No. of Birds	*Males*	*Females*	*Ratio Males to Females*	*Authority*	*Location*
669	277	381	74 : 100	Mendall & Aldous	Maine, New Brunswick, Nova Scotia
209	92	117	80 : 100	Greeley (1953)	Wisconsin
207	95	112	86 : 100	Blankenship	Michigan
5,169	2,119	3,050	70 : 100	Glasgow (1958)	Louisiana

TABLE 8. Adult Fall Sex Ratios of Some Galliform Birds *

Species	*Per Cent Males*	*Sample*	*References*
California Quail	58	511	Emlen, 1940
Bobwhite Quail	59	7,521	Bennitt, 1951
Ruffed Grouse	61	192	Bezdek, 1944
Sharp-tailed Grouse	63	103	Ammann, 1945
Sage Grouse	50.2	1,881	Girard, 1937

* Extracted from Table 8, Hickey (1955).

No one knows why the percentage of adult woodcock males in the fall bag is so low, but I can make reasonable speculations. Glasgow (1958) found that most males leave their wintering grounds on their spring migration three weeks before the females. They follow the receding snowline and, in fact, many reach northern breeding grounds while there is still more snow than bare ground. Spring snowstorms and sub-freezing weather probably kill more males than females. Male woodcock at that time are preoccupied with breeding and arrive in the North in poor flesh. The smaller males also are more vulnerable than females to severe freezes on the wintering grounds. Lanyon (1963) states: "Variations in the size of birds are related to a balance between such factors as the amount of food ingested, heat loss from the body, and the demands of flight. Since smaller birds lose more heat relative to their body size than larger birds do, there is a minimum size below which birds are incapable of ingesting and metabolizing sufficient amounts of food to offset this heat loss."

I believe also that the adult male, advertising his presence so conspicuously on the breeding grounds, is more subject to predation than the female. I have found the remains of several males on their singing grounds and have occasionally seen owls attack male woodcock. The remains of three woodcock found in a Great Horned Owl nest in May were determined to be males by measuring the uneaten bills.

Because the bird is polygamous the low proportion of males should not affect reproduction.

MINIMUM AND MAXIMUM BREEDING AGE

Woodcock of both sexes breed in their first spring on the breeding grounds; I recaptured three singing males which I had banded the previous year as juveniles. Martin (letter) examined females collected while incubating which were yearlings, so I consider it reasonable to assume they can also breed when ten months old. Maximum breeding age in woodcock is not known, but I have recaptured three singing males at least five years old and one at least six years old. Mendall recaptured a singing male in its seventh year. None of these birds was examined for fertile sperm. Probably as long as a bird, male or female, can complete its long migrations and actively court on the breeding grounds it is capable of reproducing.

LONGEVITY

Banding recoveries are steadily building evidence that female woodcock live longer than males, which accounts for the preponderance of adult females. Of 70 female recoveries from birds banded by Glasgow in Louisiana, 12 were killed four years or more after banding, while only one of 43 males was recovered four years or more after banding.

I examined the records of birds banded in Louisiana and on the breeding grounds, finding that of 124 recoveries of males, 65 per cent (81) occurred the first year, 20 per cent (25) the second year, and 15 per cent (18) the third year or later. In 128 female recoveries 48 per cent (63) were killed the first year after banding, 24 per cent (31) the second year, and 27 per cent (34) the third year or later.

In summary, 65 per cent of the males were recovered the first year and 35 per cent in later years, whereas 40 per cent of the females were recovered the first year and 51 per cent in later years. These differences are statistically significant. My earlier studies on survival of male woodcock (Sheldon, 1956) indicate that adult males are more likely to survive from spring to spring than juvenile males.

In my study areas where there was a stable or declining population, 24 males (56 per cent) out of 43 banded as adults in the spring of 1950 were alive in the spring of 1951. Of 124 male adults banded in 1951 at least 48 still survived one year later; but some birds were missed, so the actual survival rate probably was 53 per cent. Seventy-six males collected in different areas of their breeding range in the spring contained 52.6 per cent (40) adults (Martin, personal communication) and 47.4 per cent (36) yearlings.

Lack (1954) cited studies on adult mortality estimated by returns of banded birds in which the male shifts its breeding territories in successive years. His examples were the Pied Flycatcher (*Muscicapa hypoleuca*) in Finland, the European Robin (*Erithacus rubecula*) and the Redstart (*Phoenicurus phoenicurus*). He feels that this method of estimating survival is open to error but that the error is small. Since he and others have shown that for many species young of the year are more vulnerable during migration than adults, spring returns one year after initial capture provide a better yardstick than fall age ratios of the surviving breeding population, assuming there is a stable population.

WING COLLECTION

It is not known if the population composition reflected by the wing collection truly represents the status of the woodcock. Many more years of banding effort will be needed before we can assess the validity of this method. As the most complete information available, I shall assume the figures to be representative of woodcock productivity and mortality but recognize possible weaknesses of the technique.

Weather, length of hunting season, and other factors affect the annual woodcock kill. Sex and age ratio from such states as Maine and Michigan have remained virtually unchanged each year, but there is little consistency in these ratios from week to week. Disproportionate numbers of juveniles are killed on the Atlantic Coast. In

Connecticut, for example, immatures from the coast outnumber those from the interior two to one from a sample of 1,471 wings. In New Jersey coastal regions we find an even higher number of immatures in the wing samples. Baird and Robbins (letter) have captured a disproportionate number of immature passerine birds in large-scale bandings at different points on the Atlantic Coast.

There is some evidence that adult males are not taken in proportion to their actual numbers in the population. In analyzing the continental kill by dates, I found that, of 16,590 adults, 37 per cent of the kill between October 1 and October 21 (10,326 birds) were adult males, while between October 22 and November 10 (6,264 birds) 42 per cent were adult males. Where adult males migrate late the hunting pressure decreases, at least in the northern states, since interest in woodcock hunting wanes after October.

In my experience an immature woodcock is fully as wary and strong of flight as an adult by October. I have been unable to detect any difference in vulnerability to shooting traceable to age or sex of woodcock, although these differences occur in some waterfowl. Farner (1955) suggests that evidence seems to indicate that young of the year of another shorebird, the European Lapwing, are no more susceptible to shooting than the older birds. The sex ratio of juveniles in the hunter's bag is virtually even; it follows, therefore, that males and females are equally vulnerable to the gun. It is possible that differences in migration routes of age or sex groups may expose such birds to more hunters than others (especially where high numbers of immatures are present).

In spite of these potential weaknesses, the wing data are remarkable in that on a continental basis the immature/adult hen ratio has remained virtually constant for four years. The number of juveniles per adult hen was 1.8 in 1959, 1960 and 1962. In 1961 it was 1.9, only .1 higher. Martin (1965) adjusted the kill figures by weighing the totals in proportion to the reported kill in the different states. This is reasonable, although the kill figures of most states are rough estimates.

I have prepared a survival table which gives what I feel to be a reasonably accurate picture of the status of woodcock. My first step was to calculate, by percentages, the different age and sex groups in the hunter's bag. Eighteen and one-half per cent were adult males, 26.4 per cent juvenile males, 29.4 per cent adult females, and 25.6

per cent juvenile females. With this information I arrived at certain conclusions on fall to fall survival. Enough females of both age groups must survive to give an adult fall sex ratio of 63 males: 100 females. The annual survival rate of the entire female population would be 54 per cent (Table 9). A reasonable estimate of annual

TABLE 9. Woodcock Survival Table (mortality based on fall to fall)

	MALES			FEMALES		
	Living	*Dying*	*Mortality Rate*	*Living*	*Dying*	*Mortality Rate*
Hatching	1000	450	45%	1000	450	45%
1st Fall	550	369	67%	550	308	56%
2nd Fall	181	85	47%	242	92	38%
3rd Fall	96	45	"	150	57	"
4th Fall	51	24	"	93	35	"
5th Fall	27	13	"	58	22	"
6th Fall	14	7	"	36	14	"
7th Fall	7	3	"	22	8	"
8th Fall	4	2	"	14	5	"
9th Fall	2	1	"	9	3	"
10th Fall	1	0	"	6	2	"
11th Fall	1	1	"	4	2	"
12th Fall				2	1	"
13th Fall				1	0	"
14th Fall				1	1	"

Sum of living males 1st–11th fall	934	Sum of living females 1st–14th fall	1188
Sum of dead males 1st–11th fall	550	Sum of dead females 1st–14th fall	550
Mortality rate of all males $\frac{550}{934}$	59%	Mortality rate of all females $\frac{550}{1188}$	46%
Sum of living males 2nd–11th fall (Adults)	384	Sum of living females 2nd–14th fall (Adults)	638
Sum of dead males 2nd–11th fall	181	Sum of dead females 2nd–14th fall	242
Mortality rate of adult males $\frac{181}{384}$	47%	Mortality rate of adult females $\frac{242}{638}$	38%

Sum of all woodcock living 1st–14th fall	2122
Sum of all woodcock dead 1st–14th fall	1100
Annual mortality rate of total population $\frac{1100}{2122}$ =	52%

female juvenile mortality is 56 per cent and of annual adult female mortality 38 per cent (Table 9).

Of the 18.5 per cent of adult males in the fall sample, 9.8 per cent are at least two-year-olds (53 per cent survival) and approximately 8.7 per cent are replacements from the juveniles of the previous fall.

Carrying this information one step further, the survival of the fall juvenile males representing 26.4 per cent in the fall sample would not exceed 8.7 per cent of the succeeding fall's adult male population. Calculating by simple arithmetic, the annual survival of this sex and age group is 33 per cent (8.7/26.4). Using Mendall and Aldous's figures on clutch size and hatching success, I prepared a life equation (Table 10). It should be pointed out that the spring adult sex ratio

TABLE 10. Life Equation of a Stable Woodcock Population (100 birds)

		NUMBERS BY SEX AND AGE EACH SEASON				
Season	*Activities*	*Per Cent Mortality*	*Adult Males*	*Juvenile Males*	*Adult Females*	*Juvenile Females*
Spring	Arrive from migration		39 *		61 *	
	75%* of hens (46) hatch					
	3.9 eggs per clutch					
	producing			90		90
Summer	In residence; mortality of adults spring to fall	5%	37		58	
	Mortality chicks hatch to fall	45%		50		50
Fall and Winter	Mortality by sex and age		42%	63%	35%	54%
	Leaving		21	18	38	23
Spring	Breeding males		39			
	Breeding females				61	

* Estimated.

is unknown and summer mortality is based on a rough estimate. In short, I believe that woodcock can sustain an annual loss of all sex and age groups of 52 per cent and still maintain a stable breeding population.

DENSITY

Woodcock population densities in any locality are as unpredictable as the crests and troughs of ocean waves. The birds may be scattered, concentrated, or absent, depending upon the time of year, the prevailing weather or the condition of the habitat. There are innumerable records of spectacular concentrations of migrating or wintering woodcock. Hundreds stop each fall on Cape May, New Jersey.

Thousands funnel into the wintering grounds of Louisiana from far-flung breeding grounds between Minnesota and Nova Scotia. I believe the best time to measure density is during the breeding season when the birds are scattered, and the males advertise their breeding sites.

Mendall and Aldous counted an average of eight "pairs" for each 100 acres on a 1500-acre tract in Maine during the spring of 1939. An analysis of their findings indicates that they counted four occupied singing grounds per 100 acres and assumed that one male used each singing ground and that each male had a mate. Studies throw doubt on the reliability of such estimates. The bird is promiscuous; there is a floating group of opportunistic males unaccounted for; the spring male-female ratio is not known, although the fall ratio from hunters' bags contains more adult females than adult males. I believe that most of my Massachusetts singing grounds are frequented by at least three birds—one bird claiming the territory, one or more females, and one or more males seeking a singing territory. Therefore counts of singing males represent the number of occupied singing grounds present and not necessarily the population. I shall assume that the number of occupied singing grounds provides a density index of breeding populations. The number of singing grounds per 100 acres cited by woodcock workers is presented in Table 11.

TABLE 11. Breeding Density Index of Woodcock

No. of Occupied Singing Grounds per 100 Acres	*Year*	*Area in Acres*	*Location*	*Reference*
4	1939	1,500	Maine	Mendall and Aldous
2 *	1951	88,000	Massachusetts	Sheldon
5	1953	1,300	New Brunswick	Wright (per. comm.)
9.5	1940	950	Pennsylvania	Studholme, Beule, Norris
7.7	1950		Minnesota	Dangler & Marshall
.8 *	1952		Maryland	Stewart & Aldrich
3	1952		Wisconsin	Stewart & Aldrich

*These estimates cover entire range, whether in good woodcock habitat or not.

The density counts in Table 11 were concentrations of known woodcock habitat along roads, with the exception of my Massachusetts count and that of Maryland by Stewart and Aldrich (1952). My estimate is based on the total forested area of Quabbin Reservation, whether in woodcock habitat or not. In my counts in known woodcock habitat I found four to five singing grounds per 100 acres. Stewart and Aldrich (1952) state that there were three singing grounds per 100 acres in the best woodcock habitat in Maryland.

The woodcock's reputation as a recluse is warranted only in the spring. After the breeding season woodcock become gregarious at twilight. When, for example, I counted 43 birds coming into a one-acre field at dusk, I had no idea of the extent of the area they came from. In three isolated cases I could find a bird or three or four birds in the same spot every day throughout the summer, but I did not know how far they flew at dusk. A group of three woodcock could be found in precisely the same spot during the month of July, 1950, 100 yards from my residence in Shutesbury, Massachusetts. Each evening at dusk they flew in the direction of an alder brook 125 yards away; but I did not know whether they remained there during twilight. It is equally true in fall and winter that the gregarious habits of woodcock in migration or on the wintering grounds make density figures meaningless since there is no regularity in time and area where the birds gather in twilight.

MORTALITY

The woodcock is a hardy creature. It shows a remarkable capacity to recover from injuries that would be lethal to other birds. Many that are taken on both breeding and wintering areas in good health and flesh show scars of earlier injury. Some have missing toes, others have sustained broken legs that subsequently healed. Glasgow (letter) in Louisiana caught two birds, each of which had managed to survive the rigors of migration although handicapped by the loss of a leg.

In early April, 1953, I trapped a male with blood trickling from mouth and nostrils from what appeared to be a serious internal injury. When I recaptured him in late May he had recovered completely. Another April-trapped bird had broken a wing in the trap. The wing sloughed off later at the elbow joint, but the wound had healed before the bird was killed by a dog that broke into its cage in early July. I have recaptured a number of birds that had scalped themselves in trying to escape from traps when first caught. A month later they showed little sign of previous injury. Death from shock is not unusual in gallinaceous birds, but in processing 162 live woodcock for insecticide studies, there were no losses attributable to this cause.

Hunter Kill

Before the turn of the century when it was legal to shoot woodcock any month of the year and when market hunting was prevalent, gunners may well have taken a serious toll of woodcock. Even though over half a million birds a year are killed by sportsmen today, I believe that starvation, accidents, and similar causes kill more woodcock than shooting. As cited elsewhere in this volume, heavy gun pressure on local coverts before migration begins may cause some decimation of native woodcock. We have no evidence, however, that the continental population is noticeably decreased by hunting.

Band recoveries of 591 woodcock suggest to me that most die of causes other than the gun. Of these, 361 were shot, 76 died of other causes, and the remainder were retrapped and released. Superficially, the higher recovery from hunters indicates that shooting is the principal cause of death, but the chances of recovering bands from birds killed by hunters are much greater than from those that die of disease, accident, or predation. The hunter examines all birds that he kills. When he takes a banded bird, he usually has a personal interest in knowing its original banding site. Finding a banded woodcock killed by other factors is a long chance. Usually the remains are in an advanced state of decay, which discourages close examination by any but the scientifically minded. Birds killed by disease, freezing, or starvation usually die in thickets, and their plumage blends perfectly with the forest floor; only a small number of those killed by predators are found. In the light of these facts, the apparent statistical evidence that the gun is the major cause of woodcock deaths loses much weight.

Weather

Many woodcock die in years marked by periods of prolonged freezing in the wintering grounds (Mendall and Aldous). Wayne's account (1899) of abnormal snowstorm and cold in South Carolina during February 1899 is typical. Wayne says:

> The cold wave . . . was the severest recorded for two hundred years. On Monday, February 13, the thermometer registered 14°F. above zero, with the ground covered with snow from four to five inches deep on a level, while drifts were two feet deep. . . . On Tuesday, at 6:55 A.M., the thermometer registered 6° above zero.
>
> The woodcock arrived in countless thousands. Prior to their arrival, I had seen but two birds the entire winter. They were everywhere and were completely bewildered. Tens of thousands were killed by would-be sportsmen, and thousands were frozen to death. The great majority were so emaciated that they were practically feathers and, of course, were unable to withstand the cold. One man killed two hundred pairs in a few hours. I shot a dozen birds. Late Tuesday afternoon I easily caught several birds on the snow and put them into a thawed spot on the edge of a swift running stream in order that they would not perish, but, upon going to the place the next morning, I found one frozen. These were fearfully emaciated and could scarcely fly. Two birds were killed in Charleston in Broad Street. It will be many years before this fine bird can reestablish itself under the most favorable conditions.

Such serious freezes did not occur only in the past when woodcock were more abundant. A freeze struck Louisiana in the winter of 1939–40. McIlhenny (1940a), in speaking of woodcock, stated: "They got no protection from man, and were slaughtered by tens of thousands, with sticks—22 rifles and shotguns." In such cold-waves, many woodcock frequent open fields where they are easily found. This author's figures on the number killed, as well as those cited by Wayne, may be exaggerated, but there is little doubt that thousands were killed or died.

There are fewer records of woodcock deaths under unfavorable weather conditions during migration. Mendall and Aldous recorded extreme cold and heavy snow in Maine in April, 1940, when some woodcock died. Such weather conditions in spring undoubtedly cause woodcock mortality since the birds then are in relatively poor condition, and under the stress of breeding. Such losses have not been detected more often because, unlike the concentrations on the wintering ground, the birds are scattered over far-flung migration lanes. During fall migration the birds are in excellent physical condition, and it is unlikely that heavy fall freezes cause appreciable mortality, since the birds can move south. I suspect that sustained cold, rainy weather during the height of the hatch causes some losses of downy chicks. Mendall and Aldous found three recently hatched chicks killed by a heavy rainstorm. These authors estimated a 10 per cent mortality on chicks, which I believe is too low. If the collection of 40,000 wings in the fall is representative of the population composition, there would be a mortality between hatching and fall of 43 per cent (1.7 chick for each adult hen), assuming that 75 per cent of the hens hatch a brood. I doubt if drought causes woodcock deaths in their breeding grounds, since I found woodcock will readily feed on all types of invertebrates, other than earthworms, in the summer.

Hazards of Low Altitude Flights

Many woodcock fly into obstructions during migration or during crepuscular flights on their resident or winter grounds. Power lines are often hit and, occasionally, lighthouses or other constructions. I have picked up three woodcock under power lines in Massachusetts. Methodical search under wires during migration might well reveal many more deaths. The recoveries of banded birds include 45 dead from "flying into objects," "hit by automobiles," "accidents" or "found dead."

Predation

I doubt if predation is as serious a cause of woodcock deaths as the weather. Stuffed birds used as decoys in Massachusetts were attacked by crows (*Corvus brachyrhynchos*), Sharpshinned Hawks (*Accipiter striatus velox*), Cooper's Hawks (*Accipiter cooperii*), skunks, weasels, and bobcats. On two occasions I saw Great Horned Owls (*Bubo virginianus*) hunting near a singing ground. One owl tried to catch a male as he flushed and barely missed him. Observers in Louisiana have seen owls catching woodcock in their flights to and from winter feeding fields. A Great Horned Owl nest near the Massachusetts study areas contained the remains of three male woodcock. Pettingill observed a Great Horned Owl perched on a tree bordering a field where a male woodcock was performing. He suggested that courting males are conspicuous and probably attract predation by owls. On several occasions I have found the remains of woodcock on singing grounds.

Mendall and Aldous (1943) consider the house cat to be a predator of young birds, and some birds are brought in by cats in Louisiana. Long-eared Owls (*Asio otus*) are reported to be serious woodcock predators. The literature contains instances of various species of snakes, usually the Pilot Black Snake (*Elaphe obsoleta*), consuming young woodcock and woodcock eggs (Pettingill). In one season Liscinsky (1953) reported four pipped eggs taken by a housecat, three eggs broken up by a skunk, and an incubating hen killed by a Cooper's Hawk. Deaths from predation, as ascertained in five banding recoveries, were two by cats, two by dogs, and one by a shrike (*Lanius sp.*).

Nest Destruction and Abandonment

Incubating hens often abandon their eggs when flushed from their nest by man, dog or predator early in incubation. Fire is a hazard to nests. Mendall and Aldous found evidence that the practice of burning blueberry patches in eastern Maine destroyed woodcock nests. Uncontrolled fires are not as prevalent today as formerly in many parts of the woodcock breeding range, because the public has supported forest fire suppression. In Massachusetts, at one time, most of Cape Cod and Martha's Vineyard Island were almost totally burned over each spring. Forbush (1916) commented that woodcock were rare on Cape Cod and Martha's Vineyard. It may well be that better fire control in this part of the state during the last twenty years is responsible for an increase in breeding woodcock in the Cape Cod area.

Habitat Change

When there is a decrease of breeding birds the question always arises whether the birds have actually decreased or have moved elsewhere because of habitat deterioration. Southern Louisiana is a case in point. Partly as a result of large-scale drainage, birds have decreased on the lowlands, but it is possible that many moved to the pine uplands. Nice

(1937), in her studies of Song Sparrows (*Melospiza melodia*) in Ohio, found that in the late years of her studies, when much of the rough vegetation had been destroyed, only 25 per cent of banded breeding males returned as compared with 55 per cent in the first three years of banding. In this case evidence indicated that a fair proportion had not died, but had shifted their breeding territories elsewhere. Likewise, if land management practices impair large tracts of woodcock habitat, the population of this species will reflect it. I suspect that habitat change is the main factor causing the low numbers today as compared with those reported 100 years ago.

Diseases and Parasites

There is little evidence of woodcock mortality from disease or parasitism, but the chances of finding birds dead of these causes are very slim. In Maine, Mendall and Aldous found one dead emaciated female which had not died of injury. Pathological symptoms were somewhat indefinite. The trachea contained an abnormal secretion of mucous, and the liver and lungs, calcified cyst-like nodules of obscure origin. These same authors found a few endoparasites in the digestive tracts of autopsied birds. Tapeworms (*Hymenolepsis sp.*), roundworms of the family Acuaridae, and flukes of the family Strigeidae were present. Although a good many mating birds contained a few of these parasites, none was heavily infected.

Blankenship collected two Michigan birds which exhibited a diseased condition. One bird had a fungus or tumorous growth of the breast beneath the right wing. The other had scattered tumors in the body cavity involving the intestinal tract and the peritoneum and mesenteries. Neither of these diseases was identified. The same investigator found a high percentage of Michigan birds internally parasitized. In 1954 and 1955, he examined 115 birds, 109 of which were parasitized with one or more of the three major types of internal parasites: cestodes, nematodes, and trematodes. In Blankenship's 1955 specimens, there were approximately 63 parasites per bird in the first fifteen days of October, and an average of 37 parasites later in the season. The genera and species of known woodcock parasites were reviewed by Blankenship as follows:

Tapeworms (cestodes): *Hymenolepsis*

	Anomotaenia variabilis (Rausem, 1909 and Rankin, 1946)
	Anomotaenia stentorea (Fuhrman, 1932)
Roundworms (nematodes):	*Tetrameres spiralis*
	Dispharynx sp.
	Capillaris sp.
	Filaroidae (family)
	Trichostrongyloidae (family)
Flukes (trematodes):	Strigeidae (family)
	Echinostomatidae (family)

Most were from the Strigeid group and contained such genera as *Cotylurus*, *Cardiocephalus*, *Pseudapatemoa* and *Nematotstrigeia*. The same author found an arthropod parasite (Sarcosystis).

The heavy infestation found by Blankenship in woodcock exhibited no detrimental effects on the bird. He collected one woodcock in 1955 with 509 parasites of the three major types, and another with 472 tapeworms alone. Both these birds were apparently healthy, but in 1954 he recovered one sick and heavily parasitized woodcock from a bird dog.

A recent pathological study of a group of woodcock from Amherst, Massachusetts, was completed by the Patuxent Wildlife Research Center of the U. S. Bureau of Sport Fisheries and Wildlife (Locke *et al.*, 1965). Many of these birds had bumblefoot, which was probably caused by conditions of captivity, since no bumblefoot had been observed in wild woodcock. Bacteriological examination of foot lesions isolated a *Micrococcus*. Eight of 29 woodcock autopsied contained yellowish thickenings in the thoraces and abdominal air sacs. Although bacteriological studies were not made, these symptoms resembled early aspergillosis nodules or lesions described in chronic respiratory disease of domestic poultry. The first discovery of coccidiosis, a common poultry disease, was made in woodcock during this study. Six of 31 woodcock examined were infected.

ECTOPARASITES

Blankenship found two species of lice of the order Mallophaga on Michigan woodcock. A new species of louse found in Massachusetts was described by Theresa Clay of the British Museum of Natural History (personal communication) as *Rhynonirmus parsonae.*

During the winter many woodcock in Louisiana are covered with snails (*Succinea sp.*). Members of this genus are supposed to feed on the fecal material in a bird's cloaca. On the woodcock, however, they were found all over the body; they may have been seeking warmth, and so could not be classed as true parasites.

In Massachusetts during one summer, I found a number of birds infested with hippoboscid bloodsucking flies (Sheldon, Bartlett and Shaw, 1956). The two genera found were *Ornithomyia fringillina* and *Lynchia americana*, which had previously been found on grouse and other birds but never previously on woodcock. In summary, woodcock are infested with many types of parasites, but finding healthy infested birds suggests the host seldom dies from this factor.

Chemicals

The most recent threat to woodcock is the widespread spraying of chlorinated hydrocarbon chemicals (pesticides) on their wintering, migration and summer grounds. The situation may be particularly critical in wintering areas where the birds often concentrate in large numbers. Many wintering grounds in the Gulf States have, in recent years, been subjected to heavy aerial application of heptachlor to control the fire ant. Although there is no evidence that woodcock are killed directly by spraying, secondary poisoning occurs when woodcock eat earthworms on sprayed areas. Earthworms have the capacity to accumulate a considerable amount of chlorinated hydrocarbons without ill effect. Thus they present a hazard because in winter the woodcock feeds almost exclusively on worms from open fields. The pesticide hazard may be especially critical on the wintering grounds, because loss of weight during prolonged freezing periods may cause woodcock to be more susceptible to chemical poisons.

The U. S. Bureau of Sport Fisheries and Wildlife designed a four-pronged research program to find answers to these problems. It assigned a biologist to determine the population status of woodcock. His task was to collect thousands of wings from hunters, to determine the age and sex composition of fall populations, and to analyze banding methods and spring population counts. Second, woodcock carcasses were collected throughout the bird's range for chemical analysis to determine presence and amounts of pesticides. A third project developed methods for holding captive woodcock and feeding them encapsulated dosages of various chemicals. In the fourth phase, the third project was carried a step further by treating earthworms with chemicals and feeding them to captive birds on the wintering grounds. The findings were correlated with field analyses of soil and worms from treated and untreated fields.

Woodcock carcasses from the breeding and wintering ranges were collected in 1958. Approximately 25 per cent of these birds contained heptachlor epoxide (Dewitt, personal communication). Collections continued at intervals in subsequent years. Heptachlor residues were found in a progressively higher percentage of woodcock each year and in a progressively higher amount per bird. In 1962 and 1963 heptachlor epoxide was found in 148 (84 per cent) of 176 specimens from Massachusetts, Maine, New Brunswick, Ohio, and Louisiana. Concentrations ranged from less than 0.1 to 2.8 parts per million in birds taken in the north, and from traces to 5.6 parts per million from eastern Louisiana. DDT residues were present in 87 per cent of the 169 birds analyzed for this compound (Dewitt, 1963). Insecticidal residues were also found in eggs and newly hatched chicks from several nests in New Brunswick. Quantities in the eggs were approximately 4 to 15 micrograms of DDT and 0 to 10 micrograms of heptachlor epoxide.

Wright (1960, 1965) has been studying the effect of chemical poisoning on the productivity of New Brunswick woodcock for several years. Eighty-seven per cent of New Brunswick woodcock collected in the fall contained DDT or heptachlor epoxide or both. Although no information was obtained on the effect of heptachlor epoxide on productivity, Wright (1965) found a highly significant decrease of juvenile woodcock produced in thousands of acres of northern New Brunswick which were sprayed with DDT for control of the spruce budworm. Dosages varied from 2 pounds per acre to two applications of ¼ pound per acre.

A series of experiments using caged birds was conducted at Amherst, Massachusetts, in 1959 by the Massachusetts Cooperative

Wildlife Research Unit, at Patuxent Wildlife Refuge in 1960, and in Louisiana in 1961 by W. H. Stickel (Stickel *et al.* 1965a and 1965b). The response of woodcock to chemical toxicants fed in capsules varied markedly depending on condition of birds as shown by weight changes since capture. Birds which had lost weight since capture in midsummer nearly all died at heptachlor dosage levels well below those at which most birds survived in late summer and early fall when the birds had begun to accumulate fat and equalled or exceeded their weight at capture. Birds in good condition could scarcely be killed with massive doses of DDT, but succumbed to spaced oral doses when placed on starvation rations. Dieldrin appeared to be more toxic than heptachlor. Many birds excreted heptachlor doses at extremely rapid rates.

After these two series of experiments, a more realistic test was made in Louisiana (Stickel *et al.* 1965a). An abstract of the findings of this critical study follows.

> The effects on woodcock of eating heptachlor-contaminated earthworms was studied experimentally in a series of feeding trials in Louisiana in the winter of 1960–61.
>
> Six of 12 woodcock fed worms which had been contaminated at an average of 2.86 ppm of heptachlor epoxide died within 35 days; 4 more had died by the 53rd day, when the other two were killed for analysis. Worms from areas in Louisiana treated with 2 pounds of heptachlor per acre often contain more than 3 ppm of heptachlor epoxide. Eleven of 12 woodcock fed worms contaminated at an average of 0.65 ppm survived the full 60 days of the experiment (one died on the 45th day, apparently from other causes). All 11 untreated birds survived. Survivors were kept on ¼ rations of untreated worms for 11 days. Two woodcock, untreated previously, died during this starvation period. Five previously treated died; two observed in spasms at death contained 5.9 and 7.2 ppm heptachlor epoxide in their tissues, suggesting that the previous contaminated diet increased mortality, even though the difference between two of nine dying and five of nine dying is not statistically significant. Surviving starved birds given an unrestricted supply of treated or untreated worms for 5 days survived and gained weight. Residues accumulated in their tissues in this time approached levels in birds that died of heptachlor poisoning.
>
> Residues in tissues of birds with different treatment histories sug-

> gested residue loss at a rate of approximately 2.8 per cent per day. Toxicant absorption was estimated to be in the approximate range of 16 to 20 per cent. Residues in birds fed worms containing 0.65 ppm heptachlor epoxide were in the same general magnitude as those in field-caught birds, suggesting a similar average contamination of food supply.
>
> Weights and weight changes did not differ significantly among untreated birds and those receiving the lower level of toxicant. Among birds on ¼ rations, the percentage of weight that could be lost without danger seemed to be near 20 per cent.
>
> Woodcock ate 18 to 208 grams of worms per day (av. 121 grams), representing 11 to 143 per cent (av. 77 per cent) of their body weights. Birds ate contaminated and uncontaminated food in essentially equivalent amounts. Symptoms of heptachlor poisoning differed considerably between birds.

Dr. Fant Martin (1961, 1962, 1963, 1965), in analyzing productivity of woodcock from the continental wing collection from 1959 to 1962 (30,000 wings), found no significant difference in number of juveniles per adult hen for four years. There were some significant differences between productivity in certain areas but these varied each year and are believed due to factors other than mortality. Singing ground counts of courting males also have shown no drop in numbers.

In short, there is no evidence to date that aerial spraying is causing a decimation of the continental woodcock population. But results should be accepted with caution. Insufficient time has elapsed to be complacent about the status of woodcock. The harmful effects of chemicals may be accumulative over several years and may not be immediately apparent. Those responsible for managing this upland game bird are to be commended for recognizing this potential danger and for their comprehensive research.

One disquieting result is that many hunters are worried about eating woodcock that may contain heptachlor residues. Whether there is a danger in eating contaminated birds is a problem for the Department of Public Health and not the Fish and Wildlife Service. My personal view is that the danger to health is not serious if woodcock are eaten in moderation.

CHAPTER XI

European Woodcock

If an American woodcock hunter bagged a European Woodcock (*Scolopax rusticola*)—it has appeared accidentally on the North American continent—he would be amazed to pick up a bird twice as large as our native woodcock. I have found both biologists and sportsmen curious about this large cousin of our American Woodcock; superficially, it closely resembles the American genus; it is similarly specialized in its various adaptations with a life history almost identical. There are, however, differences in its physical characteristics and behavior.

HISTORY

The European Woodcock has a long history and was a table delicacy at least as early as the sixteenth century, when large numbers were netted or snared. Netting was so prevalent that the expression "cock shut" time became a synonym for twilight (Mouseley, 1935). Shakespeare, in *King Richard II*, writes:

> Thomas the Earl of Surrey and himself,
> much about cockshut time, went through the army.

Ben Johnson refers to cockshut time as follows:

> Mistress, this is only spite—
> For you would not yesternight
> Kiss him in the cockshut light.

In the Duke of Buckingham's *Household Cookbook*, 1507, eight woodcock were said to cost eightpence. Later, in the nineteenth century, large "bags" were often made. The largest bag recorded was at Ashford, Lord Adelawn's estate in County Galway, when a party of eight guns killed 508 'cock in one day in January 1895.

PHYSICAL CHARACTERISTICS

Plumage

The general color of the woodcock's plumage is russet brown, the crown with broad black bars, the rest of the upperparts are variegated, richly pencilled and marbled with black; the mantle, back and wings are rufous, the neck gray. The American bird has much more strongly marked longitudinal stripes down the back (Mouseley, 1935). The most distinct plumage coloration separating the two birds is that the underparts of the European bird are *narrowly barred* with brown on a pale buff background and vermiculated with dusky brown (Figure 42).

Figure 42. American and European woodcock

Bannerman (1961) cited two English workers who concluded young of the year could be separated from adults by the clearly notched outer web of the first primary; in the adult there is a uniform whitish edge to this feather. The most prominent plumage characteristic separating the two woodcock are the three outer primaries which are very narrow only in the American form (Figure 43).

Weight

Alexander (1934–35) found the average weight of 653 European Woodcock to be about 12 ounces with some individuals exceeding 18 ounces. Most workers say the sexes are indistinguishable, whereas the American female is heavier, has a longer bill and broader outer primaries than the male. Witherby *et al.* (1943) give wing measurements of male birds as 186–205 mm. and bills 68–79 mm. Female measurements are almost identical, with wings 184–208 and bills 67–80 mm.

Bill

The conformation of the head, bill and eyes is identical to the American bird except for a somewhat larger size. Slater (1898) describes the bill: "If the horny epidermis be removed, a number of small pits of hexagonal shape will be seen in the bone at the end of the bill, remotely suggesting an incipient honeycomb. In each of these pits a minute fibril of the olfactory nerve has its termination, and by this means, when the bill is thrust into the soft, wet soil, the slightest wriggle of the least living creature is instantly telegraphed to the woodcock's sensibility." Slater's reference to the nerves in the bill as "olfactory" is open to question especially in his conclusion

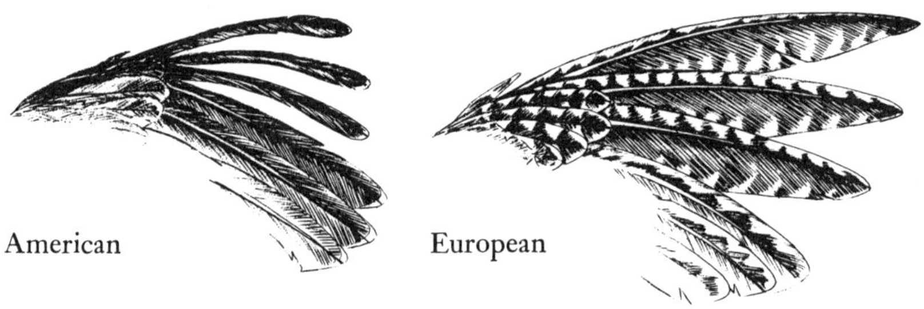

Figure 43. American and European woodcock wings

that the motion of animal life in the soil is "telegraphed" to the woodcock senses. I suspect that the bill is tactile and the sense of smell plays no part in finding food.

RANGE

Breeding Range

The European Woodcock is far more cosmopolitan in its range than our native bird. It breeds in all the British Isles. In Europe it breeds from the northern limit of forests in Norway, Finland, Sweden, and Russia south to the Caucasus, Crimea, Rumania, Bulgaria, Yugoslavia, northern Italy, and the Pyrenees. The breeding range in Asia includes Siberia, the Tian Shan Mountains, northern Mongolia, Manchuria, Japan, Iraq, Persia, the Caspian Sea area, Transcaucasia, Black Sea, Asia Minor, Syria, the valley of the Tigris, Palestine, northern Sinai, and the Nile Delta (in Africa). It is resident in the Azores, Madeira, and the western Canary Islands.

Winter Range (Bannerman, 1961)

Winter range depends a great deal on weather. Many nonmigratory birds winter in the British Isles (Alexander, 1934–35). When the weather is mild the continental birds are not driven south. In severe winters woodcock migrate to the valleys of southern mountain ranges, islands of the Mediterranean, Tunisia, Algeria, and Morocco as far south as the Great and Middle Atlas Mountains. Birds are apparently nonmigratory in the Canary Islands, possibly also in Japan. Woodcock are common in Cyprus, Corsica, Sardinia, and have been recorded in Malta. They occur casually in the Faroes, Spitzbergen, Greenland, and North America. Cooke (1912) cites records of its occurrence in Loudon County, Virginia; Chester County, Pennsylvania; Shrewsbury, New Jersey; Newport, Rhode Island; Chambly, Quebec; and St. John, Newfoundland.

BREEDING

Courtship Flights

Steinfatt's (1938) description of the spring courtship flight (referred to as "roding") of male European Woodcock during the breeding season is the most detailed account I have found. Precise descriptions are either lacking or very rare. At twilight in the evening and morning the male flies over a route described as usually triangular in shape. Three flights over this route appear the commonest; the first flight, several meters above the treetops, is the fastest; the second flight is just above the canopy, and the third and slowest flight is usually below treetop level following any natural flight lanes such as roads or other openings in the woods. During flight the male utters a guttural-like note likened to *quorr quor quor quor* immediately followed by a high note described by Steinfatt as *pssiep*.

I could find no evidence that the male alights on the ground at the same spot after each flight like the American bird. According to Hoffmann (1867), the female lures the male to the ground with a special note referred to as clucking or similar to the male *pssiep*. Steinfatt cites the description of a Swedish worker, Pay (1937), who states: "The female is not entirely without sound; it commands a piping note which to be sure is weaker and more melodious than the sound of the male." Steinfatt has shot females making the *pssiep* note, but it is not clear whether this note attracts the male or not. At times "several" males have been observed on the ground with one female (Steinfatt). Hoffmann (1862) describes the behavior of the male on the ground as follows: "The male walks around the female with ruffled feathers. In so doing it lets the wing hang down and spreads its tail out in a fan."

Copulation

Steinfatt has cited a Polish ornithologist, W. Puchalski of Lemberg, who witnessed copulation near a nest with an incomplete clutch.

> In East Poland on April 10, 1937, I sat near a woodcock nest which contained an incomplete clutch of 2 eggs. The female started to brood them as a male flew over in a courting flight and alighted 3 meters from the nest. The female emitted *cackling* sounds on the nest, arose and approached the male with hanging wings and with the tail fanned upwards towards the male. She then squatted on the ground where she was mated by the male.

Could the fanning tail ascribed by several observers to the male of the American woodcock on the courting field conceivably be a female employing the conspicuous white feathers under the tail as a lure to the male?

None of the European workers has observed the male incubating or sharing in caring for the young. It is described as a polygamous bird.

Seasonal Duration

It is of greatest interest that the European bird has two peaks of singing activity, in April and early May and in July. Steinfatt states:

> The spring mating reaches its climax in the last days of April or the first days of May; the summer mating reaches its climax in the last days of June or first days of July. From mid-May to mid-June the mating is relatively weak without stopping completely. . . . Between the climaxes of the spring and summer mating are two months, a time span which, given a breeding period lasting from 22 to 23 days, surely suffices to lead the young woodcock of the first brood to independence.

Forty-five per cent of 210 chicks banded in England and Ireland were banded in June and July, and 55 per cent in April and May. Twenty-nine per cent of them were banded in July (Alexander 1934–35). This appears to be further evidence of a second brood which Steinfatt states occurs in Prussia as well. Pay (1937) in Sweden states the woodcock has only one brood a year. Six hundred and seventy-five chicks were banded in Scotland, of which only 10 per cent were banded after July 1. It is possible that in the northern breeding range most woodcock have but one brood. There is no agreement among British workers on the number of broods. In the woodcock inquiry conducted by Alexander (1934–35) in the British Isles, 39 respondents considered the woodcock double-brooded, and 36 considered it single-brooded. Sixty-four other respondents considered the woodcock must rear at least two broods since nests or young have been found from March until September. I found no record of a hen marked so as to be recognized at different times of the season as having had two broods. In my opinion, the best evidence that some woodcock have two broods is the second peak in courtship flights in July which does not occur with the American woodcock.

Nest, Eggs and Young

The nest, number of eggs, and incubation period are the same as those of the American bird. Two hundred and thirty-eight nests of 330 found in the British Isles had 4 eggs. Sixty-four had 3 eggs, 21 had 5, 6 nests had 2 eggs and one had 6. Ticehurst (1931) states that the European woodcock female is unique amongst "waders" in having an incubation patch, a bare spot on the breast. The nest is a cup-shaped depression in the woods, usually near the edge; nests are found in the open heath very rarely. Favorite nesting areas in Scotland and elsewhere are in young plantations of pine where there are still openings between the trees. Many nests are at the base of trees as in America.

Steinfatt (1938) set up a blind one meter from a woodcock nest in Prussia. With the help of another worker, he watched the hen around the clock from early incubation until hatching. The hen left the nest at dusk and dawn each day for approximately half an hour, presumably to feed. Steinfatt built a fence of netting one foot high in a circle 24 meters in diameter. The woodcock continued to leave the nest at regular intervals when the eggs were pipped. When the young hatched she stayed constantly on the nest or in its vicinity. Steinfatt's account of her behavior is of particular interest:

> Till 4 A.M. the woodcock sat quietly on the young ones. Then it got up and remained standing for a moment at the edge of the nest. It went 20 or 30 steps from the nest and remained there 2 minutes. It then returned to the nest and pushed the eggshells with the beak from the depression of the nest to the edge while the young ones emitted low *psieh* sounds. Standing at the edge of the nest, it suddenly emitted *ug-ug-ug* which silenced the young. . . . It returned and brooded the young until 5:30 A.M. It then arose, grabbed with the beak one of the 4 young ones and carried it off walking 1½ meters from the nest where it again laid it down on the ground. As long as it held the young one in its beak, the chick emitted faint *psieh* sounds but remained quiet when deposited by the hen. She hurried back again to the nest and carried the other 3 one at a time to the first one where she brooded them.
>
> At 6:15 A.M. the woodcock arose again and, in the same manner, carried the young ones 2½ meters farther off. It then returned to the nest and pushed the eggshells back into the nest depression.

Following this the hen tried to call the chicks from outside the barrier. Steinfatt eventually opened the netting and she led the young off.

Once Steinfatt picked up one of the chicks and the hen "lifted itself above the ground ½ a meter, let the legs hang, pressed the tail downward and emitted jarring calls sounding like *pschaerr, pschaer* as it flew in a labored slow flight leading me away." While calling the chicks the hen gave a subdued vibrating *a a a a a* note.

Carrying Young

Most general ornithological books in Europe and America describe the hen purposely carrying young while flying. Mendall and Aldous never observed it in banding scores of broods. Pettingill concluded that if it occurred it must be accidental. Every European publication I have consulted describes this habit as if there were no doubt about it. Specific information from the British Isles was gathered by Alexander (1934–35), who conducted a woodcock inquiry by questionnaire. Three hundred respondents answered this question. About half said they had never observed a hen carrying young, while the others said they had. One hundred of the latter supported their observations by full particulars. Alexander quotes a retired keeper living in Cheshire, who was formerly keeper on an estate in Staffordshire, who gave the following description: "One day I was out with my master and his sons, and a woodcock got up. One of the boys immediately fired and I reproved him . . . saying he had blown its wing off. The spaniel retrieved the woodcock but it had both wings on. We went to the place where the wing had seemed to fall, and there was a young woodcock, dead but warm."

Respondents to Alexander's questionnaire were not unanimous in describing the methods woodcock use to carry their young. Ninety-seven said between the thighs next to the breast, 38 said in the feet, seven said on the back. Many of them described the tail as supporting the young bird, which suspiciously suggests that many of these observations were simply of the decoy flight of a hen woodcock flushed from her brood. Alexander (1934–35) reported on other species observed carrying young. In one case a "water rail" removed three chicks from a nest one at a time, and photographs of the incident were made. Lapwings, redshanks, snipe, and Golden Plover (*Pluvialis d. dominica*) have also been reported as carrying young on occasion (Alexander, 1934–35).

The evidence is conflicting but cannot be completely dismissed. It is certainly rare, and may be accidental, but could be a physical evolutionary adaptation for survival and not necessarily a sign of intelligence. Steinfatt's observation of a hen carrying young with her beak a few yards from her nest seems irrefutable in view of the meticulous details this ornithologist recorded about the nest under observation.

FOOD

The European Woodcock relies on earthworms and differs little from the American bird in its feeding habits. Witherby (1943) includes in its food other than earthworms: "insects" (coleoptera and their larvae, orthoptera [*Forticula*], larvae of lepidoptera). Grains of maize were found in one stomach, mussels (*Mytilus*) and small crustacea are also eaten. Seton Gordon (1915) mentions the consumption of berries, which has not been recorded for the American bird. He states: "When the blackberries have ripened the woodcock betake themselves to the hillsides and consume great quantities of the fruit."

Most European observers describe the woodcock as a bird which rests in secluded spots during the day but flies to regular feeding areas in the twilight. Selby (1833) states: "Unless disturbed, they remain quietly at roost upon the ground the whole day, but as soon as the sun is wholly below the horizon, they are in full activity, and, taking flight nearly at the same instant, leave the woods and cover for the adjoining meadow, or open land, over which they disperse themselves and are fully engaged in search of food the whole night."

HABITAT

Witherby *et al.* (1943) describe woodcock habitat as follows:

> Chiefly fragments of open woodlands and coverts in British Isles especially moist oak woods, birch, larch, spruce and Scots pine, with open glades or rides and good cover of bracken, brambles and bushes, especially evergreens, such as holly, laurel or rhododendrons. In case of conifers young plantations are largely frequented. Also rough ground with more scattered trees and/or scrub oak, birch, hazel, holly, etc. in association with ground cover of type indicated. Woods with swampy hollows and overgrown ditches are often frequented, but require dry ground to rest in by day, and flights at dusk to feed on marshes, boggy ground, and spongy places, about springs and rills. In breeding-season seclusion and freedom from disturbance appear important. In autumn before onset of cold weather is found largely on moorlands and hill country, sheltering in tall heather and bracken. In some districts breeding-stock appears to resort in similar manner to higher ground after nesting and in Ireland also breeds on bracken- and furze clad hill-

sides and even treeless marine islands. In very severe weather recorded as seeking food amongst weedy rocks on sea-shore.

MIGRATION

Schenk (1924) believed there were three migration routes in western Europe:

1. The Western Coastal Route, beginning on the eastern side of the Gulf of Finland and following the continental coasts to southwestern France.
2. The West Scandinavian Route, passing over the coast of Jutland and thereafter coinciding with Route 1.
3. The Irish-English Route, leading from the west coast of Ireland to Cornwall and then joining Routes 1 and 2 either via the south coast of England to Belgium or directly across the channel to Brittany.

Alexander (1934–35) believed Schenk's Route 2 was imaginary, since many band recoveries indicate large numbers of woodcock arrive in a wave from Scandinavia across the North Sea to the east coasts of the British Isles. Due to the wide distribution of woodcock in the Eastern Hemisphere information on migration routes is extremely fragmentary. Birds from northern Europe appear commonly in Spain, Portugal, and North Africa during the winter.

Weather plays a major role in influencing woodcock migration. In mild winters many birds from Norway and Sweden winter in the British Isles. Schenk (1924) says the dates of arrival of woodcock in Hungary range from the beginning of March to early April. When he compared yearly variations of metereological conditions and woodcock flights he found that every arrival of a large flock of birds was correlated to a barometric depression in the Northwest, especially in the British Isles. Meteorological maps of Europe disclose similar conditions, year after year, when there is a large woodcock migration. Similarly, many southern migrations of the American Woodcock occur during cold, strong northwest winds.

POPULATIONS

Alexander (1934–35) summarized reports on woodcock population trends for the last 100 years (1834–1934) from 18 areas in England,

Scotland and Ireland. Two counties reported slight decreases due primarily to habitat changes which include drainage, breaking up large wooded estates, housing developments, and many conifer woodlands growing up to forests too mature to be attractive to woodcock. Eight counties in England and Scotland and all of Ireland reported no change.

The only more recent woodcock population reports are from Denmark, which is a crossroads for woodcock migrating to and from Sweden to southern England, western France, Spain, Holland and Belgium, from the Baltic countries to Germany and France, and for breeding birds from Denmark to western France and Spain (Schack *et al.*, 1959). From 1941 to 1950 the average annual kill was 17,700 and from 1951 to 1959 it was 22,200.

Lack (1942) has worked out a few figures on population dynamics based on banding studies of 203 juveniles. His conclusions are that immature woodcock in their first year have a life expectancy of 1.7 years as compared with 2.2 years for adults. About 37 per cent of the adults die each year and must be replaced by first-year birds (to maintain a stable population). Of every 100 juveniles alive on August 1, about 44 survive to breed the next year; hence 37 first-year breeding birds correspond with 84 alive on the previous August 1. Lack did not distinguish sexes.

HUNTING

Woodcock hunting in the British Isles and Europe is better known and more popular than in America, due in part to liberal seasons and large bags. Schack *et al.* (1959) state: "The spring shooting of woodcocks, which has been practiced from ancient days, is much favored among sportsmen. The pursuit of the beautiful, mysterious bird in the spring with its poetical impression on the human mind means a much needed recreation to many sportsmen after a long winter, gives cause for a close examination of the shooting grounds, and a much valued opportunity to train the young dog. . . ."

An American woodcock hunter, E. L. Lewis (1851), also shot the European Woodcock in England: ". . . 'cockshooting' is highly prized and, in fact, is considered by many as the 'fox hunting' of shooting; while all lovers of the dog and gun regard a good day's woodcock shooting as the very *ne plus ultra* of sport. . . ." Barry

(1871), writing about woodcock hunting in England, states: "This sport has been compared in excitement to the chase properly so called; to flush and knock over a woodcock is thought by many who are experienced alike in hound, horse and gun, to be as pleasant a feat as to ride across country and come in well at the death."

There is little uniformity in the hunting seasons, and I found no mention of bag limits for woodcock in Europe. Most countries permit spring shooting. Shooting seasons for 1959 vary:

Denmark	September 24–April 7
Norway	August 21–March 1
Sweden	August 21–December 31
West Germany	September 1–April 5
Belgium	September 1–March 31
Holland	October 1–January 31
Luxembourg	August 28–April 15
France	September 4–January 6
Austria	September 1–April 15
Great Britain	September 30–February 1
Spain	September 11–March 1
Italy	August 28–March 31
East Bloc Countries	"Till early April"

Schack *et al.* (1959) justified spring shooting of woodcock by calling attention to its similarity to fall shooting of ungulates during their breeding season. In Germany there are different regulations for hunting woodcock in spring and fall (U. S. Army manual, 1959). "Pass" shooting is allowed in spring. This is not "pass" shooting in our sense of the word since it refers to what we call "dusking" woodcock or shooting them at dusk. The theory in Germany is that most of the woodcock shot by this method are males on their courtship flights. The Germans stress the importance of retrieving dogs held by leash on the spring shooting grounds since some woodcock are killed "after dark." In fall woodcock are hunted as in America, on foot with a pointing dog or spaniel.

The best area in the British Isles for hunting woodcock in modern times is in Ireland in October, where large numbers of migrants from the rest of Great Britain and Scandinavia concentrate. Much shooting in England is still confined to estates where there is no bag limit, but in other parts of Europe the sport is indulged in by many.

Epilogue

It is far from my intent to imply that we know all there is to know about woodcock. I have drawn certain conclusions based on recent advancements in unravelling the life of this fine little game bird. If this volume stimulates further research and more heated discussions among sportsmen sitting before their hunting lodge fires in October, it will have served its purpose. As I have pointed out in many places in this book, more work must be done to confirm or refute assumptions that I have made.

Hochbaum (1955) has said: "With wildlife research it is just as important to know where to go as to examine the ground already covered."

Today less than one per cent of the funds spent on research on such game as waterfowl is expended to study woodcock. It is a difficult species to study and should present a challenge to biologists and administrators. To manage this bird in a responsible way, more funds and manpower are required.

As this book goes to press I would like to call attention to what I judge to be fruitful avenues of investigation. The major question is the status of the woodcock population. Is it declining, increasing, or holding its own? Is it threatened by burgeoning human populations accompanied by far-reaching changes in the land?

Although we have developed methods that give us more confidence in our annual inventories of woodcock, there are still unknowns awaiting the curious scientist and solving them will increase the validity of our population estimates. These questions concern production and mortality. For example, we do not know the extent of the spring and summer mortality of birds returning each year to their breeding grounds. By combining spring netting techniques that I have recently developed, and summer netting before the molt with Martin's method of recognizing yearlings (subadults) it would be possible to catch large numbers of males and a smaller number of females to determine what percentage are yearlings. Knowledge of the age ratio of either sex at all times of year is one of the keys in assessing annual production and population trends. We do not know the effect of shooting; this could be partially determined by requiring the purchase of a woodcock stamp followed up by questionnaires to a sample of woodcock hunters—the method used to estimate waterfowl kill. Extensive banding is a requisite to interpret the population composition of the fall kill as revealed in the annual wing collection.

The true meaning of our annual count of singing males is a subject of dispute. We do not even know the percentage of hens which bring off a brood successfully. At what stage of incubation will a hen renest if her eggs are destroyed? More careful analysis of all types of habitat and the effect of land changes on woodcock living quarters is needed; inventory of suitable habitat may be more important than attempting to count birds. These are but a few of the questions only research can answer. I have concluded that at present the woodcock is holding its own, but this is no reason to relax vigilance, and we should continually strive to improve our methods.

Finally, I deplore the tendency of many administrators to lump the woodcock with those species of secondary importance to recreation. I can understand why they do so, since the opinions of those charged with the conservation of wild game animals inevitably reflect the pressures of the majority of sportsmen. It is probable that in most states rabbit hunters, wildfowlers, pheasant or deer hunters outnumber woodcock hunters. In the case of Michigan, however, there has been a steady increase in the number of woodcock hunters, who now equal or outnumber wildfowl hunters (Figure 57, Appendix II). For the many thousands of sportsmen who derive their greatest recreation from hunting birds in the northern uplands with well-trained dogs, woodcock are not of secondary importance, especially during years when Ruffed Grouse are scarce in the North, when we depend largely upon woodcock for sport. Only enlightened management will enable the woodcock population to withstand the increased hunting pressure of the future.

Appendices

APPENDIX I

Trapping and Netting

DECOY TRAPPING

The major feature of the Massachusetts decoy trap is its automatic action, which makes it possible for one operator to set several traps at different singing grounds the same evening (Figure 44). The action is extremely fast. In addition, the pan is large enough to capture birds on blind sets when no decoy is present. One advantage of a blind set is that females are often captured.

The mechanical parts, specifications, size, and appearance are shown in accompanying figures. The measurements need not be exactly as indicated.

Traps used by the Massachusetts Cooperative Wildlife Research Unit captured 812 woodcock during six seasons of operation, 1950–1956. At least 75 were captured in blind traps set close to traps holding decoys. A large number of records indicated that most birds sufficiently interested in a decoy could be repeatedly caught. The behavior of males in the presence of decoys was similar in most respects to the account of Norris, Beule and Studholme (1940).

If one defines a trap night as one or more traps set on one singing field baited with a decoy on the evening flight, but left blind for the dawn flight, trap night success was approximately one capture per four trap nights. Although exact figures could be calculated for success, these would be misleading, since many variables which are not measureable affect trapping success. These include weather, location of trap site, and individual skill. In addition, some males present on the singing field fail to be attracted by a decoy. In 1951, for example, 100 individual birds were trapped in 314 trap nights. In 1952 and 1953, many traps were set on former sites without first ascertaining whether birds were present. A skillful trapper operating only two traps had a much higher trap night success than another man running twelve traps, although the latter captured many more birds.

In addition to woodcocks, these traps took cottontail rabbits, a snowshoe hare, a skunk, three crows, three Cooper's hawks, two sharp-shinned hawks, several robins, whip-poor-wills, and small birds. Most mammalian captives escaped by chewing holes in the nets.

Any small machine shop can reproduce the trap successfully for approximately $20.00.

Trapping Techniques

The general procedure followed in Massachusetts was to set traps in the late afternoon, spend the singing period spotting new sites, tend traps and remove decoys after dark, leaving all traps set blind for the dawn flight. Relatively few casualties resulted from not tending dawn flight traps until late the following afternoon. In Massachusetts we trapped on all nights from late March until late May.

Birds were trapped on fields varying from several acres to alighting areas only a few feet in circumference. The smaller the singing field, the better the chances are of capturing a bird, especially on blind traps.

One man can efficiently run as many as fifteen traps. In cases where a bird was captured at a site and no new singing fields located, the trap was left on the same site for additional evenings. Many singing fields yielded several different birds during one season. In one site, five different males and a female were caught on six successive nights. Some successful traps were taken up and placed as blind sets in singing fields where a trap with a decoy had failed to catch the resident male on several successive nights. Certain birds were caught only after seven or eight trap nights.

Traps should be painted and thoroughly tested before setting them in the field. Mechanical failures, although rare, were traceable in most cases to care-

Figure 44. Author setting decoy trap (Photo by Richard Borden)

less setting. The commonest failure was to have the jaws close and spring open sufficiently to allow the bird to escape. This can have several causes, all of which can be remedied by careful trap testing and adjustment. A simple clip for holding the trap in a set position is of great convenience, because the traps are far more easily transported and handled in the field when flat. Such a clip was made with a heavy piece of wire with two open hoops clipping over the main spring trip arm (4) and passing under the backbone bar (1) (Figures 45, 46.) It is held in place by the pressure of the main spring trip arm. The trap can be thoroughly camouflaged and the clip not removed until the final setting, when the main spring trip arm is depressed sufficiently to make the clip easily removable. In order of importance, instructions for setting these traps are as follows:

1. Pinpoint the alighting spot most frequently used by a singing bird. Such spots can often be recognized in daylight by searching for the location where there is the largest deposit of droppings.
2. Camouflage the trap sufficiently to make it virtually invisible, but not so heavily as to interfere with the action. All grass, leaves, etc., should be broken up in one's hands and lightly scattered on the trap. Before the final setting, smooth the camouflage on the pan as flat as possible.
3. Place the decoy on the pan facing into the wind.
4. If sufficient traps are available, best success will be realized by making "tandem" sets—one trap with a decoy and a second blind trap nearby on an alternate lighting spot.
5. The trap should be set very lightly and can be tested by holding trip spring and depressing pan.
6. If there are several alighting spots in a field, put sticks or branches on those areas where the traps are not set, thus often guiding an alighting bird to the trapped spots.
7. A very erratic flying bird with several widely scattered alighting sites is seldom worth attempting to trap. Most birds which have been observed to alight at least three times in the same spot during an evening performance can be trapped.
8. If a trap fails to capture a bird after three nights, it should be respotted by a competent observer. Often a careful search of the area will reveal another alighting spot well advertised by fresh droppings. Traps should be shifted to the new location which may be 50 to 100 feet away.
9. Leave traps set blind for dawn flights. One hundred and thirty birds were thus captured in three trapping seasons.

Mounted woodcocks were used as decoys for the first three years. These were mounted in several poses. All took birds at one time or another. Such decoys should be stuffed large, and it makes little difference whether the legs are present or not. Those with legs were probably more visible to performing males, but squatting models were equally successful. Those mounted with the head high and the bill pointing forward seemed most attractive to males. In 1953, a dozen plastic decoys were tried. These were made out of a rubberized material prepared for the writer at the American Museum of Natural History. Several methods of painting were tried. The most crudely painted ones were as successful as those carefully painted. These decoys are very

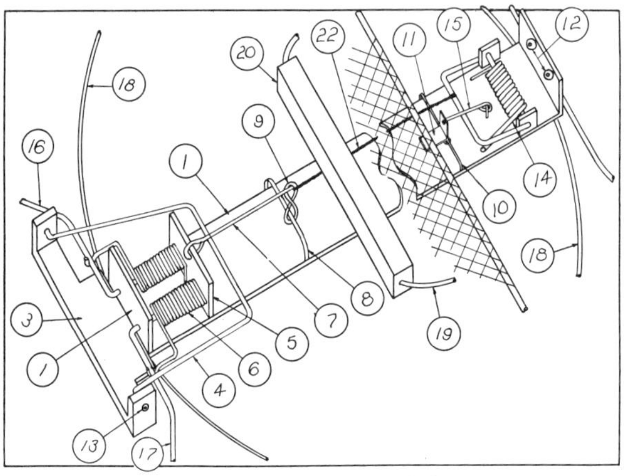

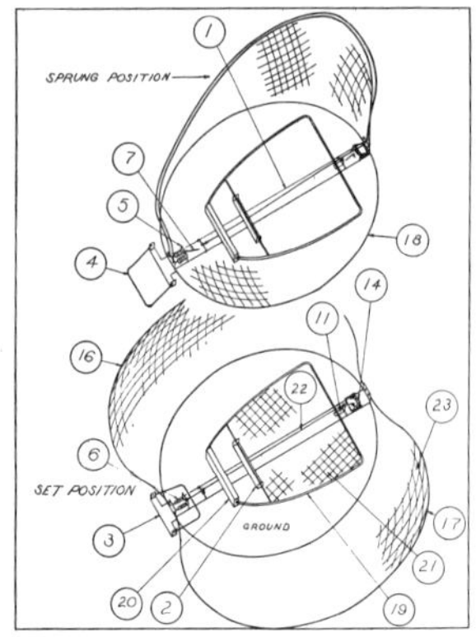

Figures 45 and 46. Decoy trap

1. Backbone bar—mild steel 1/8″ x 1-1/4″ x 22″ long (turned up 1″ at each end—holes drilled as shown).
2. Fulcrum bar-mild steel—1/8″ x 1/2″ x 6-1/2″—turned up at ends 1/2″ and depth of notches = 1/4″—width 5/32″.
3. Trip arm support welded on backbone bar—1/8″ x 1″ x 6-1/2″—cut as shown —ends turned up 1″ with 7/32″ hole.
4. Mainspring trip arm—3/16″ spring steel x 11″ long—bent as shown.
5. Bearing—mild steel 1/8″ x 1-1/4″ x 1″ deep with two 7/32″ holes drilled as shown.
6. Two main springs—1 with left turn and other with right turn—outside diam. = 11/16″ diam. of wire = 5/32″ and number of coils = 7—bent as shown.
7. Main spring trigger—3/32″ x 2-3/4″ long.
8. Main trigger trip bearing—mild steel 1/8″ x 1-1/2″ long welded on as shown.
9. Main trigger trip—figure 8—3/32″ wire.
10. Trip pan bearing—1/8″ x 1-1/2″—mild steel (brazed on).
11. Trip pan—mild steel—26 gauge—1/2″ wide x 2″ long shaped as shown.
12. 4—3/16″ burrs.
13. 2—1/8″ burrs as shown.
14. Trip pan spring—Victor mouse trap spring and arm.
15. Victor mouse trap trigger.
16. One jaw—56″—3/16″ spring steel.
17. One jaw—54″ to close inside other jaw—3/16″ spring steel.
18. Main bottom hoop support 5′ 4″ x 3/16″ spring steel.
19. Pan frame—1/8″ cold roll steel x 50″ bent as shown.
20. 2 lb. lead counter balance of main pan—diam. thickness = 13/16″, length = 9-1/2″ as drawn. A simple substitute for the square end bar is an iron pipe of the same length filled with lead.
21. Main pan mesh—heavy gauge steel lathe.
22. Picture wire trip cord.
23. 1/4″ square mesh, 1/2″ stretched, 20/6 cotton sein netting which should be dyed. To fit trap, it must be cut in egg shape and attached between jaws (16 & 17) and bottom hoop support (18). Must be very slack.

durable and appeared as effective as the feathered models. They can be left out overnight, whereas the real skin decoys quickly deteriorate in wet weather and are often destroyed by mice. In 1953, experimental trapping was conducted after the singing season by setting blind traps in favorite feeding areas. Fifteen birds were captured in these, and the method is promising if sufficiently concentrated feeding areas can be found.

MIST NETTING

Erecting Nets

The best nets for woodcock are 12 meter 4-tier Japanese nylon thrush nets, supplied by E. A. Bergstrom, editor of *Bird-Banding* (Figure 47).

After experimenting with bamboo and hardwood poles, the most satisfactory type of pole proved to be aluminum television aerial poles which come in ten-foot sections 1-inch inside diameter. They are light, sturdy, and do not rust or deteriorate when exposed to weather. To be effective for woodcock, the nets must be a minimum of 10 feet high. We put two 10-foot sections of television poles together so all our nets were 20 feet in height. Four-foot ¾-inch (outside diameter) pipes were driven 2 feet in the ground at each pole site. Each pole was placed on a pipe (Figure 48). No height was lost by having to drive a pole in the ground, and the poles were solid. Two 30-foot sets of nets were erected experimentally, but were unsuccessful since these nets were backlighted by evening sky and visible to most woodcock approaching the netting sites. In most cases, each 20-foot pole required one guy line. Four or five sets of nets were often set in series so that one pole supported two sets of nets. Birds captured above our reach were taken out with an eight-foot stepladder, but we soon found it just as rapid to lower the nets to extract a bird.

When we covered a vertical area from about 21 inches above the ground to 20 feet, at least three 4-tier nets were used. To avoid gaps between these nets, which are inevitable in 12 meter nets even when the horizontal strings are taut, the top net was first tied to the poles and the top tier of the next net was tied to overlap the bottom tier of the net above it.

The loops of each net were tied to heavy rubber bands which in turn were tied to rings of 2-inch inside diameter which slipped up and down the poles (Figures 48 and 49). To catch woodcocks, the pockets must be deep enough to drop at least half the vertical depth of the tier immediately below. This vertical distance was positively controlled by the space between the rings. The latter were tied on one vertical string, each spaced a specific distance (depending on type of net) from the ring next to it (Figures 47, 48). A heavier string was tied to the top ring and run through an eye bolt or pulley at the top of the pole and down to the ground. The operator pulled the nets up much as one pulls a flag up a flagpole. The bottom three loops of the lowest net were attached directly to the poles by rubber bands. After a netting period, these loops were pushed together vertically to bunch the net about 4 feet above the ground. After completion of netting, the rings supporting the higher nets are run down and supported by the lowest net, thus leaving all nets collapsed.

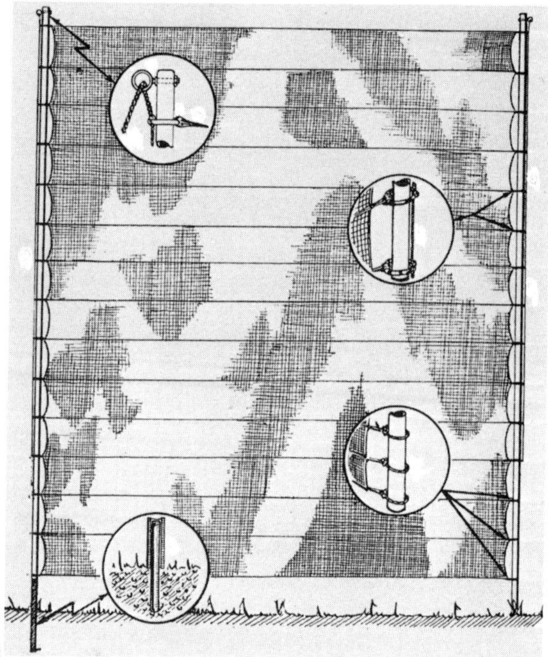

Figure 48. Mist net in position

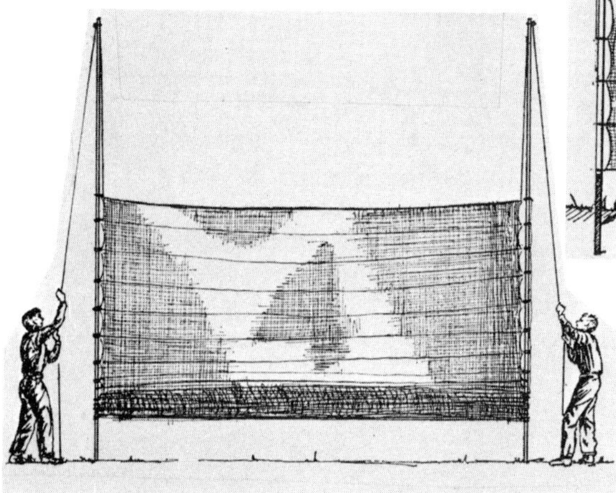

Figure 47. Hoisting mist net

Placement of Nets

Three separate sites were worked. They varied in size from ½ acre to 2½ acres. Placement of the nets proved the key to successful captures. The smallest field (Area 1) of only ½ acre was ringed with nets so it was virtually impossible for a bird to come in and escape without hitting a net. The field was narrow and long. Nets ringed the perimeter and several were placed at right angles across the width of the field. Eight sets of nets sufficed to effectively cover the field. Area 2 was about two acres in size. Placement of nets was determined by observation of the flight paths of incoming birds. Four sets of nets were set in a solid bank across the width of the field, and separate sets covered the usual flight paths of the birds. Ten complete sets proved successful. Area 3 was about 2½ acres in size and was covered effectively by banks of nets set in a broad pattern. Eight sets were required. The open end was bordered by relatively tall plantation pines providing a high flight lane seldom used except by birds flying too high to hit the nets. Woodcock which are netted usually fly into the field just over the tops of the trees or in any break between tree clumps. In some instances, the flight paths were channelized by cutting small trees on the field edge, thus creating a natural approach route.

APPENDIX I

Locating Netting Fields

All netting fields were near the center of spring breeding grounds, and all were used as singing grounds by males during the mating season. They varied widely in vegetation. Two were on high, dry hills, and the third in a relatively lower area. The ground was very dry, and trees immediately surrounding the perimeter were seldom more than 20 to 30 feet in height (Figures 49, 50). Fields can be found only by constant evening observation.

Figure 49. Singing and summer netting field, Area 1

Figure 50. Singing and summer netting field, Area 2

Duration of Netting

The most successful netting months were June, immediately following the breeding season, until mid-July. Birds have been captured all summer and as late as October 10, but in late July, August and September, evening flights are erratic. Of the 675 birds netted, over 500 were captured before August 1.

Length of Evening Activity

Nets were erected half an hour after sunset on clear days and slightly earlier on overcast days. Woodcock activity begins at the same light intensity as spring courtship performances begin. From the time the first bird was seen, all activity ceased within a maximum of 15 minutes. Usually the duration of activity was slightly more than five minutes. Nets left all night until after dawn failed to capture birds.

Effect of Weather

Windy evenings yielded few birds. There appeared to be less activity and often the birds detected the movement of the nets. Several years of weather records failed to reveal any positive correlation between weather conditions and bird activity. Cool evenings and temperatures between 55° and 65° often yielded a good catch, and just as often were unproductive. Extremely hot days followed by late afternoon thundershowers and a windless, humid evening constituted consistently successful conditions.

Number of Birds Seen or Captured per Evening

Birds observed over the netting fields varied from one or two to over twenty. Many passed over the field without attempting to alight. After mid-June until mid-July, one of the best fields yielded an average of approximately three birds per trap night. Records of high evening catches included 19, 15, 13, 10, 8, and 7. After August 1, the catch dropped to under one bird per trap night.

Subsequent experience by Martin (1963) in Maine and by the author suggest the most efficient height of the net is 10 feet, although 20-foot nets are the best if a field is small and surrounded by high trees. As pointed out in Chapter I, two nets 8 to 10 feet high placed at right angles to each other on a singing field are highly effective in capturing courting males.

APPENDIX II

Figures and Tables

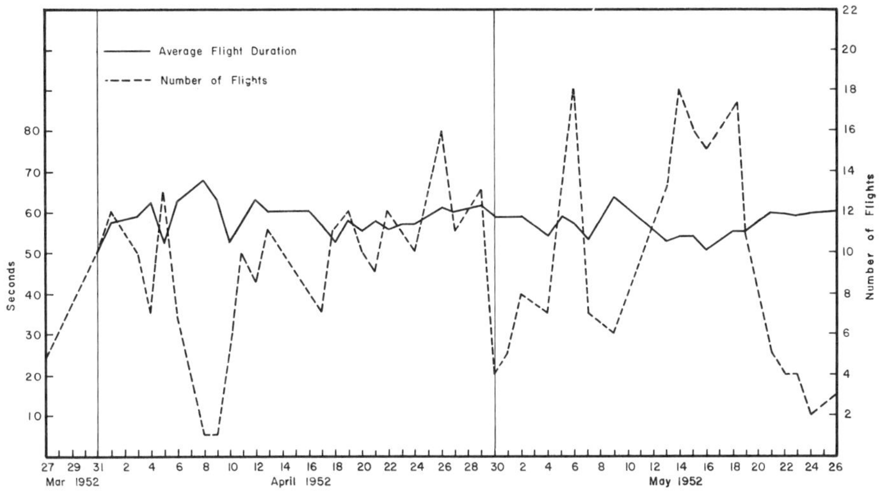

Figure 51. Courtship flights per evening of single male

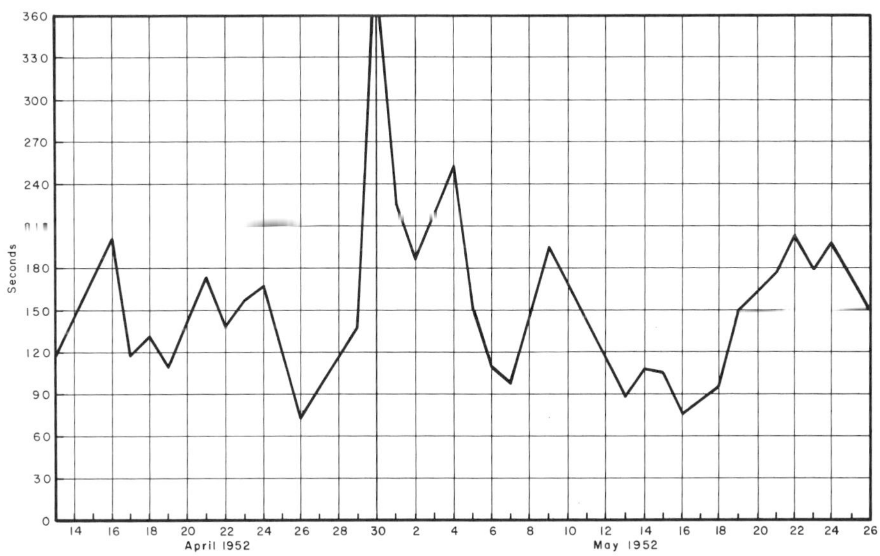

Figure 52. Average time on ground between flights of single courting male, per evening

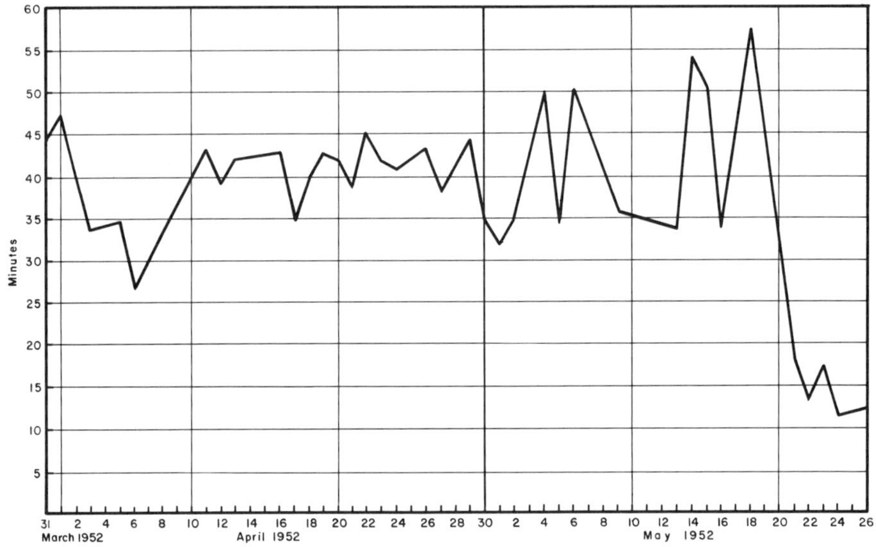

Figure 53. Duration of courtship performances of single male

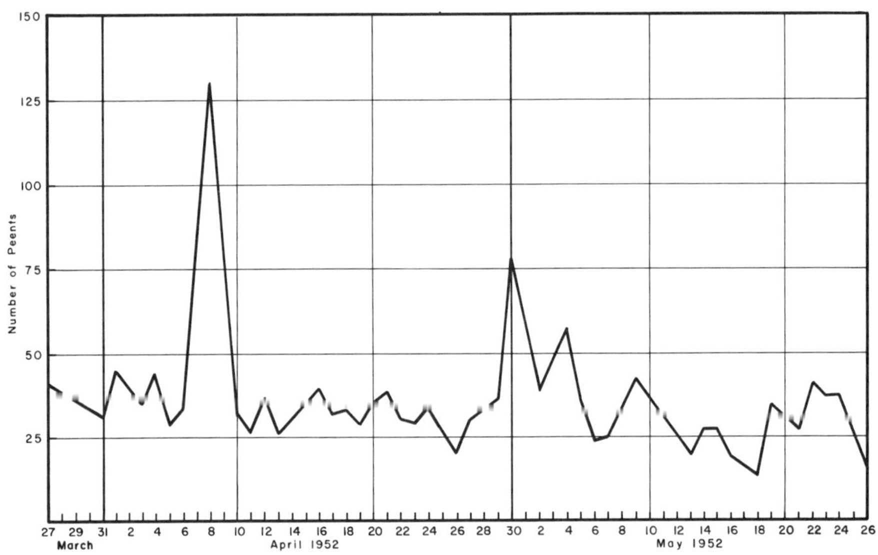

Figure 54. Average number of peents of single male between flights, per evening

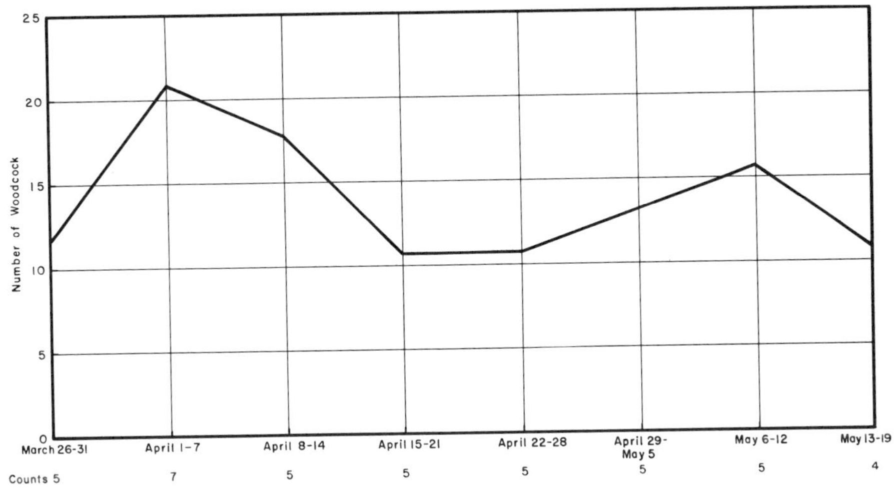

Figure 55. Mean census counts in Quabbin Reservation, 1951 and 1952

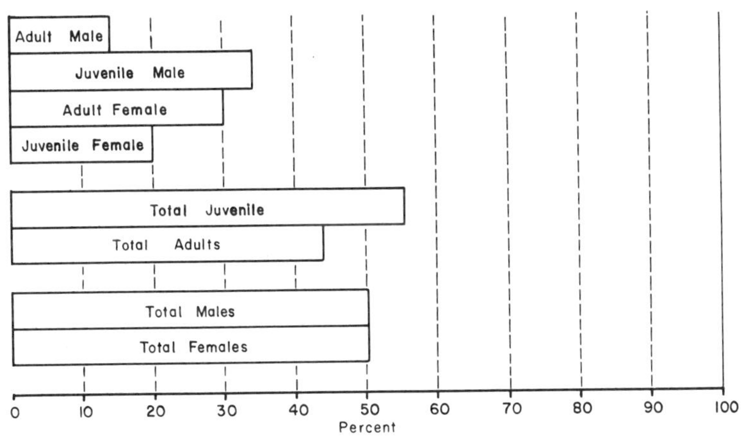

Figure 56. Sex and age ratios of 483 woodcock captured in Mass., summers 1955–59

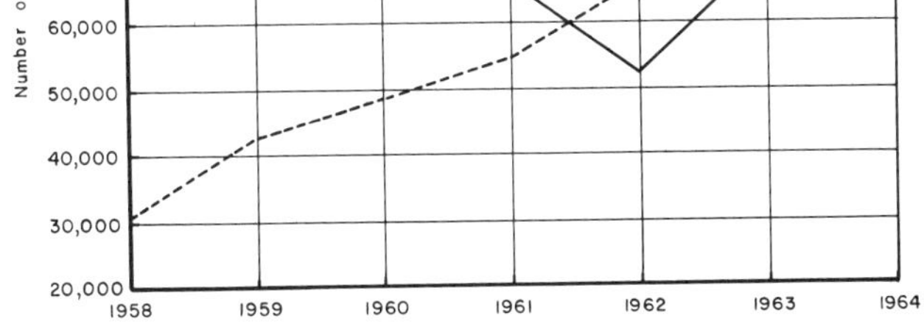

Figure 57. Numbers of wildfowl and woodcock hunters in Michigan

TABLE A. Molt Pattern of Thirteen Woodcock Collected in Massachusetts, 1955 *

Date *Age*	*Aug. 19* *Juv.*	*Aug. 19* *Juv.*	*Aug. 24* *Juv.*	*Sept. 11* *Juv.*	*Aug. 19* *Adult*	*Aug. 20* *Adult*	*Aug. 26* *Adult*	*Sept. 8* *Adult*	*Sept. 10* *Adult*	*Aug. 18* *Adult* **	*Sept. 5* *Adult*	*Aug. 26* *?*	*Aug. 26* *?*
Frons				x		x	x	x	x		x	x	x
Crown				x			x	x	x		x	x	x
Nape	x		x	x			x	x	x		x	x	x
Lores			x	x		x	x	x	x		x	x	x
Auriculars	x		x	x		x	x	x	x		x	x	x
Chin					x	x		x	x		x	x	
Neck						x	x						
Throat	x		x	x	x	x	x	x	x		x	x	x
Breast	x	x	x	x	x		x	x	x		x	x	x
Back & rump	x		x	x		x	x	x	x		x	x	x
Sides & flank	x		x	x	x	x	x	x	x		x	x	x
Abdomen	x		x	x		x	x	x	x		x	x	x
Legs	x		x	x		x	x		x			x	x
Rectrices				x					x		x	x	
Tail coverts				x	x	x		x	x		x	x	x
Wing lining			x			x							x
Scapulars			x										
Upper wing coverts			x			x							
Secondaries					x								
Primaries										x	x		

* Development of new feathers indicated by symbol X.
** Only primaries examined.

TABLE B. Summer Weights of 152 Woodcock Caught in Central Massachusetts June 13–August 28, 1956 and 1957

	Sex	*Weight Range, Grams* 1956	1957	*Average* 1956	1957	*No. of Birds* 1956	1957
Adult	M	127–162	132–165	145.9	149.5	19	12
Juvenile	M	117–149	126–152	140.4	141.1	30	19
Adult	F	163–202	162–216	182.9	189.5	27	21
Juvenile	F	151–192	152–184	168.8	166.9	16	8

TABLE C. Changes in Summer Weights (Grams) of Nine Juvenile Woodcock

Sex	*Date of First Capture*	*Weight When First Caught*	*Date of Recapture*	*Weight on Recapture*	*Weight Change*
M	6/13/57	126	6/25/57	129	+3
M	6/25/57	147	7/12/57	150	+3
M	6/23/57	121	7/23/57	124	+3
M	7/6/56	139	7/30/56	142	+3
M	6/25/58	158	7/23/58	155	−3
M	6/29/58	135	7/23/58	145	+10
F	6/19/57	154	7/7/57	171	+17
F	6/26/57	173	7/12/57	187	+14
F	6/13/57	184	7/1/57	184	0

TABLE D. Weight Increase of 179 Fall Woodcock in Nova Scotia (Tufts, 1940)

Date		*Males* *No.*	*Grams*	*Ounces*	*Females* *No.*	*Grams*	*Ounces*
October	1–15	12	160	5.6	13	202	7.1
October	16–31	42	175	6.0	31	216	7.5
November	1–15	33	180	6.3	48	[illegible]	7.9

TABLE E. Average Fall Weights (to Nearest Tenth) by Sexes Across the Breeding Range

Province or State	MALES *No.*	*Grams*	*Ounces*	FEMALES *No.*	*Grams*	*Ounces*
New Brunswick	57	166	5.8	75	208	7.2
Nova Scotia (Tufts, 1940)	87	176	6.1	92	219	7.6
Maine (Mendall and Aldous)	80	176	6.1	111	216	7.5
Wisconsin (Greeley, 1953)	81	170	5.9	104	210	7.3
Vermont (Sheldon)	31	163	5.8	33	199	6.9
Minnesota (Marshall)	171	166	5.8	221	212	7.4

TABLE F. Average Weights of 392 October Birds Shot in Minnesota (Marshall)

Type	*Number of Birds*	*Grams*	*Ounces*
Adult males	71	169	5.9
Juvenile males	100	164	5.7
Adult females	109	213	7.4
Juvenile females	112	212	7.3

TABLE G. Mean Weights of 185 Wisconsin Woodcock (Greeley, 1953)

Date	JUVENILE *Males*	*No.*	*Females*	*No.*	ADULT *Males*	*No.*	*Females*	*No.*
Oct. 1–15	165 ± 4.3 *	4	201 ± 3.4	9	164 ± 4.8	12	196 ± 3.5	15
Oct. 16–31	170 ± 1.7	32	214 ± 2.8	38	170 ± 1.6	33	215 ± 2.1	42
Totals	169 ± 2.1	36	212 ± 2.4	47	168 ± 1.8	45	209 ± 2.1	57

* Standard error of the mean.

TABLE H. Courtship Performances of 13 Male Woodcock in Central Massachusetts (May 8, 1961)

No. of Flights	*No. of Birds*	*Duration*	*No. of Birds*
9	1	30 min.	1
11	1	31 "	1
12	1	33 "	1
14	3	35 "	1
15	2	36 "	1
16	2	37 "	2
17	2	39 "	2
18	1	40 "	1
		42 "	1
		44 "	1
		49 "	1

APPENDIX II

TABLE I. Singing Periods of Eight Central Massachusetts Male Woodcock Observed Under Same Weather Conditions on Evening of May 13 and Dawn of May 14, 1963

DAWN			EVENING		
Starting Time	*Completion Time*	*Duration*	*Starting Time*	*Completion Time*	*Duration*
4:30	5:26	56 min.	8:11	8:42	31 min.
3:57	5:15	78 "	8:13	8:54	41 "
4:18	5:15	57 "	8:04	8:43	39 "
4:03	5:13	70 "	8:10	8:46	36 "
4:32	5:25	53 "	8:13	8:44	31 "
4:12	5:12	60 "	8:05	8:39	34 "
4:06	5:18	72 "	8:15	8:40	25 "
4:38	5:07	29 "	8:15	8:35	20 "
Average Time: 59 minutes				32 minutes	

TABLE J. Counts of Singing Male Woodcock in Greenbush, Maine, 1953 and 1954 (Claude Westfall)

1953		*1954*	
Date	*Count*	*Date*	*Count*
4/24	12	4/22	13
4/28	15	5/11	19
5/8	21	5/13	20

TABLE K. Size of 49 Singing Fields and Height of Bordering Trees in Central Massachusetts

Approximate Area in Square Feet	*Minimum Height of Border Trees*	*Number of Singing of Fields*
400– 900	7 feet	2
901– 2,500	10 "	1
2,600– 3,600	12 "	2
10,000–22,000	15 "	3
23,000–40,000	25 "	4
Over 40,000	40–60 "	37

TABLE L. Vegetation on Singing Grounds in Minnesota, Michigan, and Massachusetts

MINNESOTA (MARSHALL) (17 singing grounds)		MICHIGAN (BLANKENSHIP) (98 singing grounds)				MASSACHUSETTS (50 singing grounds)	
Species	*Frequency of Occurrence* *	*Species*	*Frequency of Occurrence*	*Species*	*Frequency of Occurrence*	*Species*	*Frequency of Occurrence*
				Herbs			
Grasses (dead)	15	Grasses	40	Milkweed	4	Grasses (predominantly *Andropogon*)	50
Strawberry	13	Goldenrod	30	Cinquefoil	4	Goldenrod	36
Bracken fern	8	Strawberry	19	Alsike clover	3	Cinquefoil	18
Canada mayflower	8	Pussy's toes	18	Bracken fern	2	Sensitive fern	15
Mosses	5	Nut grass	11	Bastard-toadflax	2	Sedges (*Carex*)	10
Wintergreen	3	Mosses	8	Groundsel	2	Aster	10
Anemone	3	Orange hawkweed	8	Crowfoot	2	Mosses	10
Sweet fern	3	Aster	8	Wintergreen	2	Strawberry	7
Lichen	2	Yarrow	7	Bulrush	2	Milkweed	6
		Sorrel	6	Red clover	1	Interrupted fern	2
		Camomile	6	Vetch	1	Queen Anne's lace	2
		Smartweed	5	Purple avens	1	Iris	2
		Lichen	5	Common thistle	1	New York fern	1
		Dandelion	4	Violet	1	Clover	1
		Wormwood	4	Wood fern	1		
		Hawkbit	4	Mint	1		

TABLE L. (con't.).

Woody Plants							
Blueberry	14	Willow	15	Jack pine	1	Conifers (white pine, red pine, spruce)	31
Willow	13	Quaking aspen	11	Bush clover	1	Meadow sweet	37
Trembling aspen	13	Meadow Sweet	6	Black ash	1	Raspberry (*Rubus*)	25
Beaked hazel	11	Black cherry	6	Common juniper	1	Blueberry	25
Speckled alder	9	Staghorn sumac	5	Buffalo berry	1	Alder	12
Raspberry (*pubus*)	8	Hawthorn	4	Bearberry	1	Juniper	10
Amelanchier	7	American elm	4	Canadian Juneberry	1	Grey birch	10
Norway pine	7	Blackberry	4	Dewberry	1	Aspen	9
Rose	6	Red osier	3	Wild raisin	1	Grey dogwood	8
Black cherry	5	Grey dogwood	2	Swamp birch	1	Sweet fern	8
Jack pine	5	Chokecherry	2	Raspberry (*pubus*)	1	Red maple	5
Black spruce	4	Currant	2	Pin cherry	1	Sheep laurel	4
Labrador tea	3	Sweet fern	2	Sugar maple	1	Grape	4
Mountain fly	3	Poison sumac	1	Witchhazel		Cherry	4
Honeysuckle	3					Oak	3
Bearberry	2					Willow	2
White spruce	2					Sumac	2
						Red osier, dogwood, rose, mayflower, elm, butternut, hickory, apple, hawthorn, checkerberry:	1 each

* By number of singing grounds.

TABLE M. Estimated and Known Hatching Dates of 41 Broods and 13 Nests in Central Massachusetts 1950–1961, Maine (Mendall & Aldous), and New Brunswick (Wright)

Hatching Dates	*Massachusetts*	*Maine*		*New Brunswick*
Before April 15	1	0		0
April 15–22	1	0		0
April 23–30	11	0		1
May 1–7	15	9	approx.	10
May 8–15	9	49	"	36
May 16–22	10	46	"	13
May 23–29	5	54	"	—
May 30–June 9	2	15	"	—
	54	173		60

TABLE N. Food Found in Stomachs of Fifteen Woodcock Collected in Summer Fields

Name	*Family*	*Order*	*No. of Stomachs*	*Estimated Per Cent Volume by Order*
Ground beetle larvae	*Carabidae*	Coleoptera	2	38.7
Weevils	*Curculionidae*	"	1	
Wireworms	*Elateridae*	"	13	
Leaf chafer larvae	(*Scarabaeidae*)	"	1	
Short-winged beetle larvae	*Staphylinidae*	"	3	
Darkling beetle larvae	*Tenebrionidae*	"	5	
March fly larvae	*Bibionidae*	Diptera	5	15.3
Long-legged fly larvae	*Dolichopidae*	"	4	
	[illegible]	"	1	
Snipe flies	*Rhagionidae*	"	8	
Stiletto fly larvae	*Thereyidae*	"	6	
Root feeding fly larvae	(*Tipulidae*)	"	1	
Noctuid moth caterpillar	*Noctuidae*	Lepidoptera	8	14.7
Pupa		"	1	
Loopers	*Geometridae*	"	1	
Earthworms		Annelida	8	30.0
Centipedes		Chilopoda	2	.3
Spiders	*Chelicerae*		2	.3
Ants		Hymenoptera	2	.3
Vegetable matter			6	.2

TABLE O. Band Returns of Woodcock Netted in Central Massachusetts During Summers, 1952–1959

	Sex	Distance Between Capture Sites in Miles	Years Between Capture	SITE OF RECAPTURE Summer Feeding Ground	Singing Ground
Adults	m	0	1	x	
	m	⅛	3		x
	m	1½	1		x
	m	1	1		x
	m	1	2	x	
	m	2	5	x	
	m	½	1		x
	m	10	1		x
	m	1⁄10	1		x
	m	0	2	x	
	m	1	1		x
	f	3	4	x	
	f	0	3	x	
	f	2½	4	x	
	f	0	2	x	
	f	0	1	x	
	f	0	2	x	
	f	1	1	x	
	f	0	2	x	
	f	0	3	x	
	f	0	1	x	
	f	0	1	x	
Juveniles	m	½	1		x
	m	3	3	x	
	m	0	3		x
	m	0	1	x	
	m	0	1	x	
	m	¼	1		x
	f	10	2	x	
	f	0	2	x	
	f	0	3	x	
	f	0	1	x	

TABLE P. Sex and Age Ratios * of 483 Woodcock Captured in Massachusetts During Summers of 1955–1959

	MALES		FEMALES					
	Adult	*Juv.*	*Adult*	*Juv.*	*Total Juv.*	*Total Adult*	*Total Males*	*Total Females*
1955	7(7) †	41(42)	27(28)	25(24)	66(66)	34(35)	49(49)	51(52)
1956	21(28)	32(44)	31(42)	16(22)	49(66)	51(70)	53(72)	47(64)
1957	20(13)	30(19)	36(23)	14(9)	44(28)	56(36)	50(32)	50(32)
1958	14(10)	42(30)	26(19)	18(13)	60(43)	40(29)	55(40)	45(32)
1959	10(11)	35(38)	28(31)	27(30)	38(42)	62(68)	44(49)	56(61)
Average	14(69)	34(166)	30(143)	20(98)	56(271)	44(212)	50(242)	50(241)

* Ratios in nearest whole percentages.
† Figures in parentheses denote numbers of birds.

TABLE Q. Band Returns in Louisiana from the Fall of 1948 to the Spring of 1956

Number of Years after Banding	*Male*	*Female*	*Sex Unknown*	*Total*
1	43	57	22	122
2	14	8	10	32
3		3	1	4
4		2		2
5		1		1
Total	57	71	33	161
Per cent	35.4	44.1	20.5	100.0

TABLE R. Overstory Density, in Per Cent, of Michigan Woodcock Covert (Blankenship, 1957)

	Feeding Area	*Loafing Area*
Trees	70	63
Shrubs	43	42
Ground cover	75	70

TABLE S. Woodcock Age and Sex Ratios, 1959–1962

Locality	Time Period	AGE RATIO Juv./Adult 1959	1960	1961	1962	SEX RATIOS Adult Male/Adult Female 1959	1960	1961	1962	Juv. Male/Juv. Female 1959	1960	1961	1962	NO. WINGS IN EACH PERIOD 1959	1960	1961	1962
Maine state-wide	Oct. 1–7	2.2	2.4	1.9	1.6	0.7	1.0	0.6	0.5	0.8	1.0	0.9	1.2	350	546	328	386
	Oct. 8–14	1.8	1.7	1.7	1.8	0.8	0.5	0.6	0.5	1.0	1.0	1.0	1.1	330	620	410	606
	Oct. 15–21	2.1	2.0	2.0	1.8	0.7	0.6	0.6	0.7	1.0	1.0	1.0	1.0	552	430	364	622
	Oct. 22–28	3.3	2.2	2.3	2.3	0.6	0.8	0.8	0.7	0.9	1.1	1.1	1.1	227	415	410	477
	Oct. 29–Nov. 10	2.5	2.8	5.1	2.5	1.0	0.5	1.9	1.0	1.0	1.4	1.8	1.0	131	112	96	344
	Overall	2.2	2.1	2.1	1.9	0.8	0.7	0.7	0.7	0.9	1.1	1.1	1.1	1590	2123	1608	2435
Michigan zones 1 & 2	Oct. 1–7	1.9	1.3	2.0	1.6	0.5	0.4	0.5	0.5	0.8	1.1	1.1	1.0	209	254	367	546
	Oct. 8–14	2.2	1.7	1.9	1.9	0.7	0.4	0.5	0.6	1.2	0.7	1.4	1.0	203	228	326	364
	Oct. 15–21	1.2	1.5	1.5	1.3	0.4	0.7	0.5	0.5	1.2	1.0	1.0	1.6	167	340	261	263
	Oct. 22–28	1.1	1.3	1.2	1.0	0.6	0.6	0.6	0.7	0.8	1.2	1.3	1.2	105	189	145	154
	Oct. 29–Nov. 9	1.1	1.3	1.0	1.2	0.9	0.6	0.6	0.7	1.6	0.5	1.5	1.8	108	78	127	81
	Overall	1.5	1.4	1.6	1.5	0.6	0.5	0.5	0.5	1.0	0.9	1.2	1.1	792	1089	1226	1408
Northwestern Lower Peninsula	Oct. 1–7	1.8	1.3	2.2	1.5	0.5	0.4	0.5	0.4	1.4	0.9	1.2	1.0	78	110	156	181
	Oct. 8–14	1.7	2.0	2.5	1.8	0.4	0.4	0.5	0.5	1.4	0.6	1.3	1.2	100	67	69	52
	Oct. 15–21	0.8	0.9	2.2	1.3	0.3	0.4	0.5	0.2	—	0.6	0.9	1.4	36	77	44	49
	Oct. 22–Nov. 9	0.7	1.5	0.7	1.5	0.6	0.7	0.4	0.8	2.1	1.3	0.8	1.2	78	82	84	87

TABLE S. (con't.). Woodcock Age and Sex Ratios, 1959–1962

		AGE RATIO				SEX RATIOS								NO. WINGS IN EACH PERIOD			
		Juv./Adult				*Adult Male/Adult Female*				*Juv. Male/Juv. Female*							
Locality	*Time Period*	*1959*	*1960*	*1961*	*1962*	*1959*	*1960*	*1961*	*1962*	*1959*	*1960*	*1961*	*1962*	*1959*	*1960*	*1961*	*1962*
	Overall	1.3	1.4	1.7	1.5	0.5	0.5	0.5	0.5	1.3	0.8	1.1	1.1	292	336	353	369
New Brunswick	Sep. 11–30	2.5	2.0	2.0	2.6	—	0.6	—	—	—	1.1	—	—	27	116	16	32
	Oct 1–10	2.8	2.0	1.6	2.1	1.0	0.6	0.8	0.8	1.0	1.1	1.6	0.9	267	357	116	191
	Oct 11–20	2.5	2.1	2.1	1.8	0.7	0.8	0.6	0.6	1.1	1.1	0.8	1.3	269	388	305	196
	Oct 21–31	3.5	2.1	3.0	1.9	1.0	0.9	1.2	0.7	1.5	1.4	1.4	1.3	193	210	110	109
	Nov. 1–10	2.9	—	—	—	0.8	—	—	—	1.6	—	—	—	42	0	0	20
	Overall	2.9	2.0	2.1	2.0	0.9	0.7	0.8	0.7	1.2	1.2	1.1	1.2	798	1071	547	544
New Hampshire	Oct 1–7	1.8	1.7	1.8	1.8	1.0	0.5	0.9	0.7	1.6	1.0	1.0	1.5	112	164	119	140
	Oct 8–14	2.1	1.9	1.0	1.5	0.7	0.6	0.5	0.7	0.8	0.7	1.0	1.1	158	134	159	241
	Oct 15–21	1.6	1.1	1.4	1.4	0.8	0.3	0.4	0.4	1.1	1.0	1.2	1.0	130	148	164	170
	Oct 22–28	1.8	2.3	1.7	1.4	0.5	0.9	0.5	0.8	0.9	0.9	1.5	1.3	87	185	132	225
	Oct 29–Nov. 9	1.9	2.5	2.1	1.4	0.6	1.3	0.7	0.9	1.0	1.5	1.6	1.2	123	52	68	195
	Overall	1.8	1.7	1.4	1.5	0.7	0.6	0.6	0.7	1.0	0.9	1.2	1.2	610	683	642	971
New Jersey state-wide	Oct 16–22	1.4	1.8	1.3	2.6	0.5	0.6	0.7	0.6	0.9	1.1	0.8	1.4	267	271	122	161
	Oct 23–29	2.1	1.2	1.8	1.9	0.3	0.5	0.4	0.4	1.0	0.8	1.1	0.8	197	307	101	269
	Oct 30–Nov. 5	3.1	4.8	2.6	2.1	0.5	0.8	0.9	0.6	1.0	1.4	1.7	0.8	230	139	102	212
	Nov. 6–25	2.6	5.1	5.3	2.3	0.4	0.7	1.2	0.7	1.1	1.0	0.8	1.2	158	128	232	250

TABLE S.(con't.).

	Overall	2.1	2.0	2.6	2.1	0.4	0.6	0.8	0.6	1.0	1.0	1.0	1.0	852	845	557	892
New York	Oct. 5–11	1.2	1.3	1.3	1.5	0.7	0.5	0.8	0.4	1.0	1.1	1.4	0.9	215	108	140	313
	Oct. 12–18	1.3	1.6	1.3	1.2	0.5	0.6	0.4	0.6	0.7	1.3	1.0	1.1	227	302	307	339
	Oct. 19–25	1.3	1.8	1.7	1.7	0.5	0.7	0.7	0.7	1.4	1.1	0.9	1.1	182	306	260	379
	Oct. 26– Nov. 1	1.4	1.2	1.3	1.8	0.7	0.7	0.8	0.8	0.9	0.8	1.1	1.3	116	188	147	272
	Nov. 12–18	1.5	1.6	1.6	1.5	0.8	1.0	1.3	0.9	1.4	1.3	1.2	1.3	109	82	197	196
	Overall	1.3	1.5	1.4	1.5	0.6	0.7	0.7	0.7	1.0	1.1	1.1	1.1	849	986	1051	1499
Pennsylvania	Oct. 13–19	1.9	1.6	1.9	1.8	0.7	0.5	0.8	0.6	0.8	1.0	1.1	1.1	216	321	414	664
	Oct. 20–26	2.0	1.6	2.0	2.0	1.0	0.7	0.7	0.7	1.2	1.1	1.6	1.4	128	251	187	296
	Oct. 27– Nov. 2	2.4	1.9	1.3	1.1	1.2	0.9	0.6	0.6	1.4	1.4	1.0	1.3	115	88	112	337
	Nov. 3–22	4.1	3.5	1.9	1.7	1.4	0.9	0.9	0.8	1.5	0.6	0.9	0.7	59	43	175	196
	Overall	2.2	1.7	1.8	1.7	0.9	0.6	0.8	0.7	1.1	1.0	1.1	1.1	518	703	888	1493
Wisconsin state-wide	Oct. 1–7	2.1	1.5	1.7	2.8	0.5	0.7	0.4	0.6	0.8	0.8	1.0	0.9	257	202	235	295
	Oct. 8–14	1.7	1.7	1.6	2.4	0.5	0.5	0.4	0.6	1.0	1.0	0.9	1.0	242	149	217	275
	Oct. 15–21	1.4	1.3	1.9	1.7	0.5	0.4	0.6	0.5	0.6	1.3	0.8	1.8	217	324	246	244
	Oct. 22–28	1.7	1.5	1.2	2.9	0.7	0.6	0.9	0.8	0.9	0.9	0.5	0.9	107	283	120	249
	Oct. 29– Nov. 9	1.1	1.5	2.1	1.8	0.6	0.9	1.0	0.6	1.0	1.0	0.5	1.1	122	142	68	120
	Overall	1.6	1.5	1.6	2.3	0.6	0.5	0.8	0.6	0.3	1.0	0.8	1.1	945	1100	886	1183

APPENDIX III

List of Plants

WOODY PLANTS

Alder (*Alnus spp.*)
Apple (*Pyrus Malus*)
Ash, black (*Fraxinus nigra*)
Aspen, large-toothed (*Populus grandidentata*)
Aspen, trembling (*Populus tremuloides*)
Bald Cypress (*Taxodium*)
Balsam fir (*Abies balsamea*)
Bearberry, common (*Arctostaphylos Uva-ursi*)
Beech, American (*Fagus grandifolia*)
Birch, canoe (*Betula papyrifera*)
Birch, gray (*Betula populifolia*)
Birch, swamp (*Betula pumila*)
Blackberry (*Rubus candensis*)
Blackberry, common (*Rubus allegheniensis*)
Blackberry, dwarf (*Rubus pubescens*)
Blueberry, black highbush (*Vaccinium atrococcum*)
Blueberry, highbush (*Vaccinium corymbosum*)
Blueberry, lowbush (*Vaccinium vacillans*)
Blueberry, low sweet (*Vaccinium angustifolium*)
Buffalo-berry (*Shepherdia argentea*)
Butternut (*Juglans cinerea*)
Button bush (*Cephalanthus occidentalis*)
Cedar, eastern red (*Juniperus virginiana*)
Cedar, northern red (*Juniperus virginiana* var. *crebra*)
Cedar, white (*Chamaecyparis thyoides*)
Cherry, black (*Prunus serotina*)
Cherry, choke (*Prunus virginiana*)
Cherry, pin (*Prunus pensylvanica*)
Chestnut (*Castanea dentata*)
Cottonwood (*Populus deltoides*)
Crab-apple (*Pyrus* spp.)
Currant (*Ribes* spp.)
Dewberry (*Rubus hispidus*)
Dewberry, northern (*Rubus flagellaris*)
Dogwood (*Cornus Drummondi*)
Dogwood, flowering (*Cornus florida*)
Dogwood, gray-twig (*Cornus racemosa*)
Dogwood, red osier (*Cornus stolonifera*)
Dogwood, silky (*Cornus Amomum*)
Elm, American (*Ulmus americana*)
Elm, water (*Planera aquatica*)
Grape (*Vitis* spp.)
Greenbrier (*Smilax* spp.)
Gum, black (*Nyssa sylvatica*)
Gum, tupelo (*Nyssa aquatica*)

Hackberry (*Celtis laevigata* and *Celtis occidentalis*)
Hardhack (*Spiraea tomentosa*)
Hawthorne (*Crataegus* spp.)
Hazelnut (*Corylus americana*)
Hazelnut, beaked (*Corylus cornuta*)
Hemlock (*Tsuga canadensis*)
Hickory (*Carya* spp.)
Holly, or possum-haw (*Ilex decidua*)
Honeysuckle (*Lonicera* spp.)
Honeysuckle, fly (*Lonicera villosa*)
June-berry (*Amelanchier* spp.)
Juniper, pasture (*Juniperus communis* var. *depressa*)
Juniper, upright pasture (*Juniperus communis*)
Labrador tea (*Ledum groenlandicum*)
Laurel, sheep (*Kalmia angustifolia*)
Locust (*Robinia Pseudo-Acacia*)
Locust, water (*Gleditsia aquatica*)
Magnolia (*Magnolia* spp.)
Maple, red (*Acer rubrum*)
Maple, sugar (*Acer saccharum*)
Mayflower (*Epigaea repens*)
Meadowsweet (*Spiraea latifolia*)
Mulberry, French (*Callicarpa americana*)
Oak, blackjack (*Quercus marilandica*)
Oak, cherrybark (*Quercus falcata* var. *pagodaefolia*)
Oak, chestnut (*Quercus Prinus*)
Oak, northern red (*Quercus rubra*)
Oak, post (*Quercus stellata*)
Oak, scrub (*Quercus illicifolia*)
Oak, Spanish (*Quercus falcata*)
Oak, water (*Quercus nigra*)
Oak, white (*Quercus alba*)
Oak, willow (*Quercus phellos*)
Pecan, bitter (*Carya aquatica*)
Peppervine (*Ampelopsis arborea* and *Ampelopsis cordata*)
Persimmon (*Diospyros virginiana*)
Pine, jack (*Pinus Banksiana*)
Pine, Loblolly (*Pinus Taeda*)
Pine, longleaf (*Pinus australis*)
Pine, pitch (*Pinus rigida*)
Pine, red (*Pinus resinosa*)
Pine, short leaf (*Pinus echinata*)
Pine, slash (*Pinus caribaea*)
Pine, white (*Pinus Strobus*)
Plum (*Prunus americana*)
Poison ivy (*Rhus radicans*)
Privet, swamp (*Forestiera acuminata*)
Raspberry, black (*Rubus occidentalis*)
Raspberry, flowering (*Rubus odoratus*)
Raspberry, red (*Rubus strigosus*)
Rose (*Rosa* spp.)
Service-berry (*Amelanchier* spp.)
Shadbush (*Amelanchier* spp.)
Snowberry, creeping (*Gaultheria hispidula*)
Spruce, black (*Picea mariana*)
Spruce, Norway (*Picea Abies*)
Spruce, red (*Picea rubens*)
Spruce, white (*Picea glauca*)
Sumac, poison (*Rhus vernix*)
Sumac, staghorn (*Rhus typhina*)
Sumac, wing-rib (*Rhus copallina*)
Supplejack (*Berchemia scandens*)
Sweet fern (*Comptonia peregina*)
Sweet gum (*Liquidambar Styraciflua*)
Sycamore (*Platanus occidentalis*)
Tree of Heaven (*Ailanthus altissima*)
Tulip tree (*Liriodendron tulipifera*)
Wax myrtle (*Myrica cerifera*)
Wild raisin (*Viburnum cassinoides*)
Willow (*Salix* spp.)
Wintergreen (*Gaultheria procumbens*)
Witch hazel (*Hamamelis virginiana*)
Yaupan (*Ilex vomitoria*)

HERBACEOUS PLANTS

Anemone (*Anemone* spp.)
Aster (*Aster* spp.)

Basil, mountain mint (*Pycnanthemum* spp.)
Blue-flag (*Iris versicolor*)
Bluestem, common (*Andropogon scoparius*)
Bluestem, pinehill (*Andropogon divergens*)
Bristlegrass (*Setaria geniculata*)
Buffalo weed (*Ambrosia trifida*)
Bulrushes (*Scirpus* spp.)
Buttercup (*Ranunculus* spp.)
Camomile, wild (*Matricaria Chamomilla*)
Cat gut (*Tephrosia* spp.)
Chocolate plant (*Melochia corchorifolia*)
Cinquefoil (*Potentilla* spp.)
Clover, alsike (*Trifolium hybridum*)
Clover, bush (*Lespedeza* spp.)
Clover, Japanese (*Lespedeza striata*)
Clover, red (*Trifolium pratense*)
Cocklebur (*Xanthium italicum*)
Coneflowers (*Rudbeckia* spp.)
Dandelion (*Taraxacum erythrospermum* and *Taraxacum officinale*)
Devil's paintbrush (*Hieracium aurantiacium*)
Fern, brake (*Pteridium aquilinum*)
Fern, interrupted (*Osmunda Claytoniana*)
Fern, New York (*Dryopteris noveboracensis*)
Fern, sensitive (*Onoclea sensibilis*)
Fern, shield (*Dryopteris* spp.)
Goldenrod (*Solidago* spp.)
Grass, Bermuda (*Cynodon dactylon*)
Grass, carpet (*Axonopus affinis*)
Grass, crab (*Digitaria sanguinalis*)
Grass, Dallis (*Paspalum dilatatum*)
Grass, headland (*Paspalum urvelli*)
Grass, Johnson (*Sorghum halepense*)
Grass, nut (*Cyperus* spp.)
Grass, panic (*Panicum* spp.)
Grass, signal (*Brachiaria platyphylla*)
Grass, smut (*Sporobulis poiretii*)
Grass, switch (*Panicum virgatum*)
Groundsel, common (*Senecio vulgaris*)
Hawkbit (*Leontodon autumnalis*)
Marsh fleabane (*Pluchea* spp.)
Millet (*Echinochloa Walteri*)
Palmetto, dwarf (*Sabal minor*)
Peppermint (*Mentha piperito*)
Purple avens (*Geum rivale*)
Pussy's toes (*Antennaria* spp.)
Queen Anne's lace (*Daucus carota*)
Ragwort (*Senecio* spp.)
Sedge (*Carex* spp.)
Sesbania (*Sesbania exaltata*)
Smartweed (*Polygonum* spp.)
Sneezeweed (*Helenium tenuifolium*)
Sorrel, green (*Rumex Acetosa*)
Sorrel, mountain (*Oxyria digyna*)
Sorrel, red (*Rumex acetosella*)
Sorrel, wood (*Oxalis* spp.)
St. Andrew's cross (*Ascyrum Hypericoides*)
St. Peter's wort (*Ascyrum stans*)
Strawberry (*Fragaria virginiana*)
Sumpweed (*Iva ciliata*)
Switch cane (*Arundinaria tecta*)
Thistle, bull (*Cirsium vulgare*)
Tie vine (*Thyella tamnifolia*)
Toadflax, bastard (*Comandra umbellata*)
Vetch (*Vicia* spp.)
Violet (*Viola* spp.)
Wintergreen, flowering (*Polygala paucifolia*)
Wintergreen, one-sided (*Pyrola secunda*)
Wintergreen, pink (*Pyrola asarifolia*)
Wintergreen, spotted (*Chimaphila maculata*)
Wormwood, sage (*Artemisia ludoviciana* and *Artemisia vulgaris*)
Yankee weed (*Eupatorium capillifolium*)
Yarrow (*Achillea Millefolium*)

APPENDIX IV

Key for Aging Woodcock from Wings

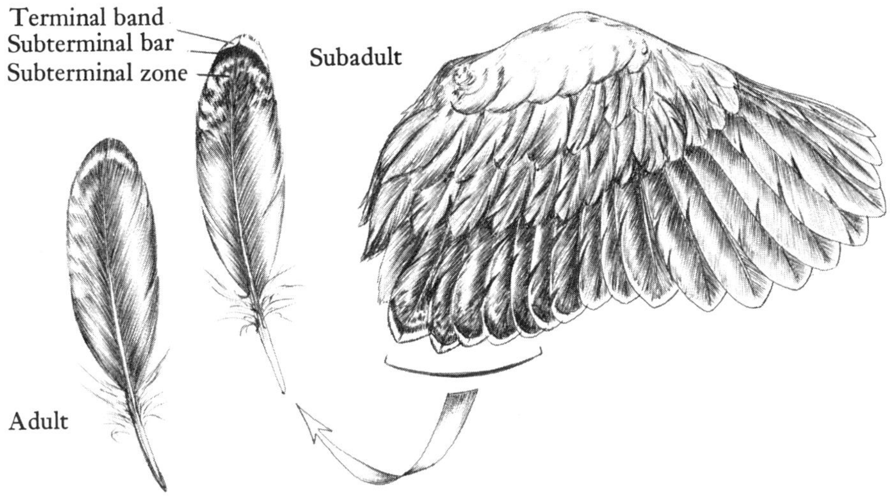

Figure 58. Aging by secondary feather patterns

January through March (when immatures are not present in the population)

I. Terminal light-colored band present on underside of tip of two or more secondaries: [1]

A. Band contrasts greatly with rest of feather . . . . Subadult (first-year adult)

B. Band not contrasting greatly with rest of feather:

1. Subterminal dark bar bordering light-colored terminal band:

(a) Bar well defined; pointed at midrib and forms generally smooth border with lighter-colored terminal band; darker or denser than rest of feather . . . . . . Subadult (first-year adult)

(b) Bar poorly defined; if discerned, distal edge forms uneven or poorly outlined border with lighter-colored terminal band . . . . Adult (older than 1 year)

2. Subterminal dark bar absent . . . . Adult (older than 1 year)

[1] When examining secondaries for presence of terminal band and subterminal bar, omit the four to six distinctly barred secondaries nearest the body and examine underside of tip of the next five feathers.

April through December (when immatures are present in the population)

- I. Primary feathers not molting and primaries and secondaries not noticeably faded and worn:
  - A. Terminal light-colored band present on underside of tip of two or more secondaries: [1]
    - 1. Band contrasts greatly with rest of feather . . . . Immature
    - 2. Band not contrasting greatly with rest of feather:
      - (a) Subterminal dark bar bordering light-colored terminal band:
        - (1) Bar well defined; pointed at midrib and forms generally smooth border with lighter-colored terminal band; darker or denser than rest of feather . . . . Immature
        - (2) Bar poorly defined; if discerned, distal edge forms uneven or poorly outlined border with lighter-colored terminal band . . . . Adult (Sept.–Dec.)
      - (b) Subterminal dark bar absent . . . . Adult (Sept.–Dec.)
  - B. Terminal light-colored band absent . . . . Adult (Sept.–Dec.)
- II. Primary feathers molting or primaries and secondaries much faded and worn:
  - A. Terminal light-colored band present on underside of tip of two or more secondaries: [1]
    - 1. Band contrasts greatly with rest of feather . . . . Subadult (first-year adult)
    - 2. Band not contrasting greatly with rest of feather:
      - (a) Subterminal dark bar bordering light-colored terminal band:
        - (1) Bar well defined; pointed at midrib and forms generally smooth border with lighter-colored terminal band; darker or denser than rest of feather . . . . Subadult (first-year adult)
        - (2) Bar poorly defined; if discerned, distal edge forms uneven or poorly outlined border with lighter-colored terminal band . . . . Adult (older than 1 year)
      - (b) Subterminal dark bar absent . . . . Adult (older than 1 year)
  - B. Terminal light-colored band absent . . . . . . Adult (older than 1 year)

Selected Bibliography

Selected Bibliography

Alexander, W. B. 1934–35. The woodcock in the British Isles. The Ibis, Vol. 87, pp. 512–550; Vol. 88, pp. 1–24, 159–169, 271–286, 427–444; Vol. 89, pp. 1–28.

Allen, Glover A. 1962 (1925). Birds and their attributes. Dover Pub., New York. 338 pp.

Ammann, G. A. 1963. Methods of capturing woodcock hens and chicks on their breeding grounds. Report No. 2414. Mich. Dept. of Conser., Game Div.

Anonymous. 1930 (1783). The sportsman's companion. Privately republished by Ernest R. Gee (200 copies), pp. 32–33.

Audubon, J. J. 1859. The birds of America. Vol. 6: 17. G. R. Lockwood & Son, New York. 456 pp.

Bagg, A. C. 1923. The Connecticut valley—a highway for bird migration. Auk, 40: 256–275.

Bannerman, D. A. 1960. Birds of the British Isles. European Woodcock. Oliver & Boyd, Edinburgh & London. Vol. 9, pp. 96–110.

Barnes, William. 1955. Pittman-Robertson wildlife research report. Ind. Dept. of Conser., 16: 92–98.

Barrows, W. B. 1912. Michigan bird life. Mich. Agr. Coll., Lansing. 822 pp.

Barry, W. 1871. Moorland and stream with notes and prose idyls on shooting and trout fishing. Tinsley Bros., London, pp. 162–173.

Bennett, L. J. and P. F. English. 1947. The 1936 fall woodcock population. Pennsylvania Game News, 17: 14–15.

Bent, A. C. 1927. Life histories of North American shorebirds (Part I). U.S. Nat. Mus. Bull. 142, U.S. Gov't. Ptg. Office. 420 pp.

Beyer, O. E., A. Allison and H. H. Kopman. 1908. List of birds of Louisiana. Auk, 25: 173–180.

Blankenship, L. H. 1957. Investigations of the American woodcock in Michigan. Mich. Dept. of Conser., Report No. 2123. 217 pp.

Bogardus, A. H. 1874. Field, cover, and trap shooting. V. B. Ford & Co., New York, p. 132.

Bowles, D. Guillermo. 1775. Introducción a la historia natural, y a la geografía física de Espana. Madrid.

Boyer, G. F. 1959. Northern distribution of woodcock in Ontario. Fish and Wildlife Mgt. Report No. 48. Ont. Dept. of Lands and Forests.

Brewster, W. 1894. Note and song flight of the woodcock. Auk, 11: 291–298.

Brodkorb, P. 1956. Pleistocene birds from Eichelberger Cave, Florida. Auk, 73: 136–137.

Brown, C. P. 1953. Woodcock management research. Final report, P-R Project 53-R. New York Conser. Dept. 31 pp.

Brown, J. L. 1964. The evolution of diversity in avian territorial systems. Wilsons Bulletin 76, No. 2, pp. 160–169.

Bump, Gardiner, Robert W. Darrow, Frank C. Edminster and Walter F. Crissey. 1947. The Ruffed Grouse. New York Conser. Dept. 915 pp.

Butler, A. W. 1898. Birds of Indiana. Ind. Dept. of Nat. Res. 1187 pp.

Cobb, Stanley. 1959. On the angle of the cerebral axis in the American Woodcock. Auk, 76 (1): 55–59.

Cooke, Wells Woodbridge. 1912. Distribution and migration of North American shore birds. U.S. Dept. Agr., Biological Survey, Bulletin No. 35, revised.

Cottam, C. 1935. Adaptability in the feeding habits of the woodcock. Wilson Bull., 46: 200.

Critcher, T. S. and T. L. Quay. 1953. The American Woodcock in North Carolina. Chat, 17: 62–68.

Cumberland, G. 1846. Vol. of bird drawings including anatomy. See Mouseley, Henry.

Dangler, E. W. and W. H. Marshall. 1950. Wodcock studies at Cloquet Forest Experiment Station, Minnesota. Flicker, 22: 1–7.

Dawson, W. L. 1903. The birds of Ohio. Wheaton Pub. Co., Columbus. 671 pp.

Dexter, R. W. 1949. Aquatic snails eaten by woodcock. Auk, 66: 88.

Duke, G. E. 1964. Study of the reliability of censuses of migrating male woodcock. Master's thesis. Mich. State Univ. of Agriculture and Applied Science.

Duvall, A. J. 1955. The juvenal plumage of the American Woodcock. Investigations of woodcock, snipe and rails in 1956. U.S. Fish and Wildlife Serv. Spec. Sci. Rept. Wildl. No. 31, pp. 43–46.

Eaton, W. H. 1910. Birds of New York. New York State Mus. Memoir 12, Pt. 1. 298 pp.

Edminster, F. C. 1954. American game birds. Chas. Scribner's Sons, New York.

Eneminger, A. B. 1954. Earthworm populations on wintering areas of the American woodcock in the vicinity of Baton Rouge, Louisiana. Unpublished Master's thesis. La. State Univ. 97 pp.

Farner, Donald S. 1955. Bird-banding in the study of population dynamics. Recent studies of avian biology, Albert Wolfson, ed., Univ. of Ill. Press. Chapter 12: pp. 397–449.

Fayle, D. C. F. 1961. Who needlessly sets foot upon a worm. The Forestry Chronicle, Vol. 37, No. 1, pp. 10–15.

Fernald, M. L. 1950. Gray's Manual of Botany, 8th ed. American Book Company, New York.

Fisher, A. K. 1902. Two vanishing game birds: the woodcock and the Wood Duck. Yearbook U.S.D.A. 1901, pp. 447–458.

Forbush, Edward H. 1916. A history of the game birds, wildfowl and shorebirds of Massachusetts and adjacent states. State Board of Agriculture, Boston. 2nd ed. XVIII: 636 pp., 36 pl., 26 figs.

———. 1925. Birds of Massachusetts and other New England states. Mass. Dept. Agr., Boston. 3 vols. illus.

Gates, G. E. 1961. Ecology of some earthworms, with special reference to seasonal activity. Am. Mid. Nat., Vol. 66, No. 1, pp. 61–86.

Glasgow, Leslie L. 1953. The American Woodcock (*Philohela minor*) in Louisiana. (Mimeo.) Trans. S. E. Game Conf., Chattanooga, Tenn. 16 pp.

———. 1956. A method of measuring wintering woodcock populations on nocturnal feeding sites in Louisiana. U.S. Fish and Wildlife Serv. Spec. Sci. Rept. Wildl. No. 31: 7–10.

———.1957a. The wintering woodcock population on nocturnal feeding sites in Louisiana 1955–56. U.S. Fish and Wildlife Serv. Spec. Sci. Rept. Wildl. No. 34: 51–55.

———. 1957b. Preliminary analysis of woodcock band returns. U.S. Fish and Wildlife Serv. Spec. Sci. Rept. Wildl. No. 34: 9–10.

———. 1958. Contributions to the knowledge of the ecology of the American Woodcock, *Philohela minor*, on the wintering range in Louisiana. Ph.D. thesis. Texas A. & M. 153 pp.

Gordon, Seton. 1915. Hill birds of Scotland. Arnold, London.

Goudy, William H. 1960. Factors affecting woodcock spring population indexes in southern Michigan. Mich. Dept. of Conser. Game Div. Report 2281.

Grant, William C., Jr. 1955. Studies on moisture relationship in earthworms. Ecol., Vol. 36, No. 3, pp. 400–407.

Greeley, Frederick. 1953. Sex and age studies in fall-shot woodcock (*Philohela minor*) from southern Wisconsin. Jour. Wildl. Mgt. 17: 29–32.

Gregory, J. U. 1893. A South Carolina woodcock flight. Forest and Stream, 40: 74.

Grinnell, G. B. 1922. Woodcock carrying its young. Auk, 39: 563–564.

Griscom, Ludlow. 1949. The birds of Concord. Harvard Univ. Press, Cambridge. 840 pp.

Gunn. W. W. H. 1951. The woodcock project. (Mimeo.) Ont. Dept. Lands and Forests, Div. of Research, Wildlife Sec.

Handley, W. R. C. 1954. Mull and mor information in relation to forest soils. Forestry Can., Bull. No. 23. Her Majesty's Stationery Office, London.

Harris, Bruce K. 1965. More specimen records of birds unusual in New Mexico. Auk, 82: 648–650.

Harris, S. W. 1953. Sex ratio studies of fall-shot woodcock (*Philohela minor*) from Pine County, Minnesota. Unpublished manuscript, Univ. of Minn.

Hawley, Ralph Chipman and Austin F. Hawes. 1912. Forestry in New England. John Wiley & Sons, New York. 347 pp.

Herbert, H. W. (Pseudonym: Forester, Frank). 1951. Frank Forester on upland shooting, A. R. Beverly, ed. Giddings, William Morrow and Co.

Herman, Walter J. 1952. Survey of the earthworms of Lincoln Parish, Louisiana. Proc. La. Acad. of Science, Vol. XV, Aug., pp. 19–23.

Hickey, Joseph J. 1951. Some American population research on gallinaceous birds. Recent studies in avian biology, Albert Wolfson, ed. Univ. of Ill. Press. Chapter 11: pp. 326–396.

Hitchcock, A. S. 1950. Manual of the grasses of the United States. 2nd ed. U.S. Gov't. Ptg. Office.

Hochbaum, H. A. 1944. The Canvasback on a prairie marsh. The Stackpole Co., Harrisburg, pp. 1–207.

———. 1955. Preface to Prairie Ducks by Lyle K. Sowls. The Stackpole Co., Harrisburg.

Hoffman, Julius. 1867. Die Waldschnepfe; ein Monograph beitrag zur Jagdzoologie. 1. Aufl. 151S. Stuttgart; 2 aufl. Stuttgart 1887.

Holland, W. C. 1944. Physiographic divisions of the Quarternary lowlands of Louisiana. La. Acad. of Sci., 8: 11–24.

Howe, H. V., R. J. Russell, F. B. Kniffen, J. H. McQuirt and S. M. McDonald. 1938. The geology of Iberville and Ascension parishes. Dept. of Conser., La. Geol. Surv. Bull. 13. 223 pp.

Howell, A. H. 1911. Birds of Arkansas. U.S. Dept. Agr., Biological Survey Bulletin No. 38, p. 29.

Knight, J. A. 1944 and 1946. Woodcock. Alfred Knopf, New York.

Kozicky, E. L., T. A. Bancroft and P. G. Horneyer. 1954. Analysis of woodcock singing ground counts. Jour. Wildl. Mgt., Vol. 18, No. 2, pp. 259–266.

Lack, D. 1943. The age of some more British birds. Brit. Birds, 36: 214–221.

———. 1954. The natural regulation of animal numbers. Clarendon Press, Oxford, 343 pp.

Lanyon, W. E. 1963. Biology of birds. The Nat. Hist. Press for Am. Mus. of Nat. Hist., New York.

Latham, R. M. 1956. Midwinter courtship flights of woodcock in Delaware. Jour. Wildl. Mgt., 20: 209.

Le Jeune, Paul. 1635. In the Jesuit relation and allied documents. Burrows Bros. Co., Cleveland [1897]. Vol. VI, pp. 95–317.

Leopold, Aldo and Alfred E. Eynon. 1961. Avian daybreak and evening song in relation to time and light intensity. The Condor, 63: No. 4, pp. 269–293.

Lewis, E. J. 1851. Hints to sportsmen, containing notes on shooting. Lea and Blanchard, Philadelphia, pp. 94–114.

———. 1863. The American sportsman. Philadelphia, p. 178.

Ligon, J. D. 1963. A Pleistocene avifauna from Haile, Florida. Unpublished Master's thesis. Univ. of Fla.

Lincoln, F. C. 1951. A set of 12 eggs of the woodcock. Auk, 68: 376.

Liscinsky, S. A. and William J. Bailey. 1955. A modified shorebird trap for capturing woodcock and grouse. Jour. Wildl. Mgt., 19: 405–408.

———. 1962. Pittman-Robertson research and development projects. Pa. Game Comm., 9: 48–54.

———. 1964. The Pennsylvania woodcock study. Pa. Game News, Vol. 35 (10), Oct., pp. 14–17.

———. 1964. Woodcock management in Pennsylvania. Complete final report. In manuscript.

———. 1965. The American Woodcock in Pennsylvania. Bulletin published by Pa. Game Comm., Harrisburg, Pa.

Livingston, R. B. 1964. A guide to the spring and summer flora of western New England. Hamilton I. Newell, Amherst.

Locke, Louis N., William H. Stickel and Shirley A. Geis. 1965. Some diseases and parasites of captive woodcocks. Jour. Wildl. Mgt. 29 (1): 156–161.

Lowery, G. H., Jr. 1955. Louisiana birds. La. State Univ. Press, Baton Rouge. 556 pp.

Lynch, J. J. 1951. Woodcock in south Louisiana during the freeze of January–February, 1951. U.S. Fish and Wildlife Serv. Spec. Sci. Rept. Wildl. No. 14, pp. 3–9.

MacConnell, William P. and Robert S. Bond. 1961. Application of herbicides with mist blowers: a promising method of releasing conifers. Journal of Forestry, 59 (6): 427–432.

Marshall, William H. 1958. Woodcock singing grounds at the Cloquet Experimental Forest, 1947–1956. Trans. N. Amer. Wildl. Conf., pp. 296–305.

Martin, F. W. 1961. Woodcock status report, 1961. U.S. Fish and Wildlife Serv. Spec. Sci. Rept. Wildl. No. 58. 29 pp.

———. 1962. Woodcock status report, 1962. U.S. Fish and Wildlife Serv. Spec. Sci. Rept. Wildl. No. 69. 36 pp.

———. 1963. Woodcock status report, 1963. U.S. Fish and Wildlife Serv. Spec. Sci. Rept. Wildl. No. 76. 43 pp.

———. 1963. Effects of regulations on size of the woodcock kill. U.S. Fish and Wildlife Serv., Migratory Bird Populations Station. Administrative Rept. No. 39. 5 pp.

———. 1964. Summer banding of woodcock. U.S. Fish and Wildlife Serv., Migratory Bird Populations Station. Administrative Rept. No. 43. 8 pp. (with E. Clark).

———. 1964. Woodcock age and sex determination from wings. Jour. Wildl. Mgt. 28 (2): 287–293.

———. 1964. A preliminary report on results of woodcock wing collections, 1959–62. Trans. N. Amer. Wildl. Conf., Hartford, Conn. 10 pp.

———. Aelred D. Geis, and William H. Stickel. 1965. Results of woodcock wing collections, 1959 to 1962. Jour. Wildl. Mgt. 29 (1): 121–131.

Maxfield, H. K. 1961. A vegetational analysis of fifty singing grounds in central Massachusetts. Unpublished Master's thesis. Univ. of Mass.

Mayer, John. 1828. The sportsman's directory. 5th ed. Printed for Baldwin and Cradock; Longman, Reese, Orme, Brown and Green, W. Joy; and Swinborne, Walter, and Taylor. London. 204 pp.

Mayr, Ernest. 1935. Bird territory. Proc. Linnaean Soc. N.Y., pp. 45–46.

McIlhenny, E. A. 1940a. Effect of excessive cold on birds in southern Louisiana. Auk, 57: 408–410.

Mendall, H. L. and C. M. Aldous. 1943. The ecology and management of the American Woodcock. Maine Coop. Wildl. Res. Unit, Univ. of Maine. 201 pp.

Merovka, L. J. 1939. The woodcock in Louisiana. Louisiana Conser. Review, 8 (4): 10–14.

Miller, Donald R. 1957. Soil types and earthworm abundance in woodcock habitat in central Pennsylvania. Master's thesis. Dept. of Zoology and Entomology. Pa. State Univ. 69 pp.

Montgomery, J. P. 1956. Trends in farm land acreage, use and value in Louisiana, 1909–1954. La. State Univ., Dept. of Agri. Econ. Circular No. 190. 33 pp.

Mouseley, Henry. 1935. A historical review of the habits and anatomy of the woodcock. Compiled from the earliest drawings and accounts to those of the present day. Can. Field Nat., XLIX (1): 1–28.

Murchie, William R. 1958. Biology of the *Oligochaete Eisenia rosea* (Savigny) in an upland forest soil of southern Michigan. Am. Midl. Nat., 60: 113–131.

———. 1960. Biology of the *Oligochaete Bimastos zeteki* Smith and Gitting (*Lumbricidae*) in northern Michigan. Am. Midl. Nat., 64: 194–215.

Musselman, T. E. 1945. Bald eagles and woodcocks in central-western Illinois. Auk, 62: 458.

Neuhaus, Walter. 1962. On the olfactory sense of birds. Olfaction and taste. Zooterman, ed. Macmillan Co., New York, pp. 111–123.

Nice, M. M. 1937. Studies in the life history of the song sparrow. Vol. 1. Trans. Linnaean Soc. N.Y., 4: 1–247.

———. 1941. The role of territory in bird life. Am. Midl. Nat., 26 (3): 441–487.

Norris, R. T. 1941. Woodcock wintering ground studies in Louisiana, Dec. 1940–Feb. 1941. Manuscript. Report to U.S. Fish and Wildlife Serv. 45 pp.

———. J. D. Beule and A. F. Studholme. 1940. Banding woodcocks on Pennsylvania singing grounds. Jour. Wildl. Mgt., 4 (1): 8–14.

———. 1942. Woodcock wintering ground studies in the southern states, Dec. 1941–Feb. 1942. Manuscript. Report to U.S. Fish and Wildlife Serv. 18 pp.

Oberholser, H. C. 1938. The bird life of Louisiana. La. Dept. of Conser., New Orleans. Bull. 28. 834 pp.

Pay, L. M. Die Waldschnepfe. Munchen 1937 (aus dem. schweidischen ubersetzt) 101 Seiten.

Payne, Fred. 1962. Woodcock trapping techniques. (Mimeo.) Univ. of Maine.

Pettingill, O. S., Jr. 1936. The American Woodcock *Philohela minor* (Gmelin). Mem. Boston Soc. Nat. Hist., 9: 169–391.

Pitelka, F. A. 1943. Territoriality, display, and certain ecological relations of the American Woodcock. Wilson Bull., 55: 88–114.

Preble, Edward A. 1902. A biological investigation of the Hudson Bay region. N. Amer. Fauna, No. 22. USDA, pp. 47–48.

———. 1908. A biological investigation of the Athabaska-Mackenzie region. N. Amer. Fauna, No. 27. USDA, p. 317.

———. 1927. The vanishing woodcock. Nature Mag. 9: 235–240.

Reardon, J. Douglas. 1950. Woodcock utilization of improved covers in eastern Maine. Unpublished thesis. Univ. of Maine.

Reid, V. and P. Goodrum. 1952. Wintering woodcock populations in west-central Louisiana. U.S. Fish and Wildlife Serv. Spec. Sci. Rept. Wildl. No. 14, pp. 10–14.

———. 1953. Wintering woodcock populations in west-central Louisiana, 1951–52. U.S. Fish and Wildlife Serv. Spec. Sci. Rept. Wildl. No. 18: 4–12.

———. 1954. Observations of woodcock breeding in certain southeastern states. U.S. Fish and Wildlife Serv. Spec. Sci. Rept. Wildl. No. 24: 17–18.

———. 1955. Wintering woodcock populations in west-central Louisiana, 1953–54. U.S. Fish and Wildlife Serv. Spec. Sci. Rept. Wildl. No. 28: 7–11.

———. 1957. Wintering woodcock observations in west-central Louisiana, 1956. U.S. Fish and Wildlife Serv. Spec. Sci. Rept. Wildl. No. 34: 11–16.

Rice, J. H., Jr. 1924. Destruction of birds in South Carolina. Auk, 41: 171–172.

Richter, W. C. 1948. Singing ground analysis and cover preferences of the American Woodcock in central Pennsylvania. Unpublished Master's thesis. Pa. State College.

Ripley, Thomas H. 1957. The bobwhite in Massachusetts. Bull. No. 15, Mass. Div. Fish and Game, Boston.

Roberts, T. S. 1936. The birds of Minnesota. Univ. of Minn. Press. Vol. 1. 719 pp.

Rowan, William. 1931. The riddle of migration. Williams and Wilkins Co., Baltimore. 151 pp.

Russell, D. M. 1954. Annual progress report, woodcock studies, July 1, 1953 to May 31, 1954. Ky. Dept. of Fish and Wildlife Resources, Frankfort, pp. 56–74.

———. 1958. Woodcock studies in Kentucky. P-R Project W-31-R, Ky. Dept. of Fish and Wildlife Resources.

Sandys, E. and T. S. Van Dyke. 1904. Upland game birds. Macmillan, New York, p. 321.

Sargent, Charles Sprague. 1949. Manual of the trees of North America. 2nd ed. Dover Publ., New York.

Schack, K. 1959. Memorandum (mimeo.) from Dansk jagtforening, landsjagtforeningen Af 1923 and Dansk strandjagtforening. (Danish Sportsmen's Assn., The National Society of Sportsmen of 1923, and the Danish Inland Shooting Association.) English translation.

Schenk, Jakab. 1924. Az erdei szalonka vonulasa Europoban. Der zug der Waldschnepfe in Europa. Aguila 30 (1): 26–120.

———. 1925. The migration of the woodcock in Europe. Brit. Birds, 19: 34–44.

Schorger, A. W. 1929. Woodcock carrying young. Auk, 46: 232.

Selby, Prideaux John. 1833. Illustrations of British ornithology. W. H. Lizars, Edinburgh.

Sheldon, W. G. 1952. Studies on breeding woodcock in Massachusetts. Investigations of woodcock, snipe, and rails in 1951. U.S. Fish and Wildlife Ser. Spec. Sci. Rept. Wildl. No. 14, pp. 30–44.

———. 1953a. Report on woodcock investigations in Massachusetts spring 1952. Investigations of woodcock, snipe, and rails in 1952. U.S. Fish and Wildlife Serv. Spec. Sci. Rept. 1: Wildl. No. 18, pp. 28–29.

———. 1953b. Woodcock studies in Massachusetts. Trans. N. Amer. Wildl. Conf., 18: 369–377.

———. 1954. Summary of Massachusetts woodcock studies. In Investigations of woodcock, snipe, and rails in 1953. U.S. Fish and Wildlife Serv. Spec. Sci. Rept. Wildl. No. 24, pp. 35–39.

———. 1955a. Massachusetts woodcock studies—1954. Investigations of woodcock, snipe, and rails in 1954. U.S. Fish and Wildlife Serv. Spec. Sci. Rept. Wildl. No. 28, pp. 29–33.

———. 1955b. Methods of trapping woodcocks on their breeding grounds. J. Wildl. Mgt., 19 (1): 109–115.

———. 1956a. Annual survival of Massachusetts male woodcocks. J. Wildl. Mgt., 20: 420–427.

———. 1956b. Massachusetts woodcock studies. Investigations of woodcock, snipe, and rails in 1955. U.S. Fish and Wildlife Serv. Spec. Sci. Rept. Wildl. No. 31, pp. 43–46.

———. F. R. Shaw and L. M. Bartlett. 1956c. New records of the Hipposcids on woodcock. Proc. of Ent. Soc. of Wash., 58: 13.

———. F. Greeley, and John Kupa. 1958. Aging fall-shot American Woodcocks by primary wear. Jour. Wildl. Mtg., 22: 310–312.

———. 1959. Cover mapping a five million acre state from aerial photographs. Trans. N. Amer. Wildl. Conf., pp. 313–321.

———. 1960. A method of mist netting woodcocks in summer. Bird-Banding, 31: 130–135.

———. 1961. Summer crepuscular flights of American Woodcock in central Massachusetts. Wilson Bull., 73 (2): 126–139.

Slater, Henry Horrocks. 1898. British birds with their nests and eggs. Order Limicolae. F. W. Frohawk(?). London, Brumby & Clarke.

Small, John Kunkel. 1933. Manual of the Southeastern Flora. Univ. of North Carolina Press.

Smith, F. 1919. *Lumbriculidae*. Can. Arctic Exped. 1913–1918. Annelids, parasitic worms, protozoans, etc. Vol. IX; Part A, pp. 3–8, illus.

Smith, R. W. 1901. The cold wave of February 1899 in Dekalb County, Georgia. Wilson Bull., 13: 35–36.

Sperry, C. C. 1940. Food habits of a group of shorebirds: woodcock, snipe, knot and dowitcher. U.S. Biol. Surv. Wildl. Res. Bull. 1. 37 pp.

Stegeman, Leroy C. 1960. A preliminary survey of earthworms of the Tully Forest in central New York. Ecol., 41 (4): 779–782.

Steinfatt, Otto. 1938. Das Brutleban der Waldschnepfe. Jour. für Ornithologie, 86: 379–424.

Stewart, Robert E. and John Aldrich. 1951. Removal and repopulation of breeding birds in spruce-pine forest community. Auk, 68: 471–482.

Stickel, William H., Don W. Hayne, and Lucille F. Stickel. 1965a. Effects of heptachlor-contaminated earthworms on woodcocks. Jour. Wildl. Mgt. 29: 132–146.

———. Wendell E. Dodge, William G. Sheldon, James B. Dewitt and Lucille F. Stickel. 1965. Body condition and response to pesticides in woodcocks. Jour. Wildl. Mgt. 29: 147–155.

———. William G. Sheldon, and Lucille F. Stickel. 1965. Care of captive woodcocks. Jour. Wildl. Mgt. 29: 161–172.

Stone, Witmer. 1937. Bird studies at Old Cape May. Del. Orn. Club, Acad. Nat. Sciences, Phila., pp. 402–409.

Studholme, Allan T. and R. T. Norris. 1942. Breeding woodcock populations. Auk, 59: 229–233.

Thompson, Daniel Q. 1966. Unusual choice of cover by woodcock. N. Y. Fish and Game Journal, 12 (2). In press.

Thorp, T. B. 1846. The hive of "the bee hunter." D. Appleton & Co., New York. 312 pp.

Tinbergen, N. 1935. The function of sexual flight in birds; territory and the problem of the origin of territory. Bird-Banding, VIII (1): 1–8.

———. 1957. The function of territory. Bird Study, 4 (1): 14–27.

Tufts, R. W. 1940. Some studies in bill measurements and body weights of American Woodcock (*Philohela minor*). Can. Field-Nat., 54 (9): 132–134.

Twitle, R. A. 1890. Woodcock and snipe in the Middle West. Forest and Stream, 34: 407–408.

Van Pelt, A. 1951. All outdoors, Times Picayune, New Orleans. Jan. 21–27, 1951.

Walls, G. L. 1942. The vertebrate eye and its adaptive radiation. Cranbrook Inst. Sci., Bloomfield Hills. 785 pp.

Warren, B. H. 1888. Birds of Pennsylvania. Meyers Pub. Co., Harrisburg. 260 pp.

Wayne, A. T. 1899. Destruction of birds by the great cold wave of February 13 and 14, 1899. Auk, 16: 197–198.

Weeden, Robert B. 1955. Cover requirements of breeding woodcock in central Maine. Unpublished Master's thesis. Univ. of Maine.

Westfall, Clyde. 1954. Unpublished Master's thesis, Univ. of Maine.

Wetherbee, D. K. 1962. Egg teeth and shell rupture of the American Woodcock. Auk, 79: 117.

———. N. S. Wetherbee. 1961. Artificial incubation of eggs of various bird species and some attributes of neonates. Bird-Banding, 32: 141–159.

Wetmore, A. 1962. Smithsonian Publ., Vol. 145, No. 2, p. 10.

Wheeler, P. H. 1955. Forests of Louisiana, 1953–54. U. S. Forest Service Survey Release 75. Southern For. Exp. Sta., New Orleans. 64 pp.

White, C. L. and N. J. Foscue. 1943. Regional geography of Anglo-America. Prentice-Hall, Inc., Englewood. 518 pp.

White, F. B. 1933. Birds and motor cars. Auk, 50: 236.

Witherby, H. F., Rev. F. C. R. Jourdain, Norman F. Ticehurst, Bernard W. Tucker. 1940. The handbook of British birds. Vol. IV. H. F. & G. Witherby, Ltd., London, pp. 184–192.

Wolfson, A. 1945. The role of the pituitary, fat deposition and body weight in bird migration. Condor, 7: 95–125.

Wright, Bruce S. 1956. The relation of beaver to moose and woodcock. (Mimeo.) N. E. Wildl. Sta., Univ. New Brunswick. Fredericton.

———. 1960. Woodcock reproduction on DDT-sprayed areas of New Brunswick. Jour. Wildl. Mgt., 24 (4): 419–420.

———. 1965. Some effects of heptachlor and DDT on New Brunswick woodcocks. Jour. Wildl. Mgt., 29 (1): 172–185.

Yeager, W. 1947. Breeding populations of woodcock in central Pennsylvania. Unpublished Master's thesis. Pa. State College.

Index

Index

Accipiter cooperii, see Hawk, Cooper's
Accipiter striatus, see Hawk, Sharp-shinned
Acknowledgments, x–xii
Age, methods of separating adults and juveniles, 38, 203–204; wear on outer primaries, 38, coloration of secondaries, 38, 203
Aix sponsa, see Duck, Wood
Alabama, 22, 24, 26, 70
Alders, 63, 64, 68, 77, 79, 81, 82, 85, 123, 124, 125, 134, 135, 136, 138
Aldous, C., 79; *see also* Mendall and Aldous
Aldrich, J., *see* Stewart and Aldrich
Alexander, W. B., 161, 163, 165, 167, 169
Allen, G. A., 34, 35, 36, 78
Allison, A., *see* Beyer, Allison and Kopman
Amherst (Mass.), 7, 12, 27, 136
Ammann, G. A., 8, 20, 88, 102, 105, 120, 124, 130, 138
Anahynchus, see Plover, New Zealand Wrybill
Anser anser, see Goose, Gray Lag
Appalachians, 22
Apple, 124, 126
Arctic, 62
Arkansas, 22, 24, 26
Armstrong Township (Ont.), northernmost nest record, 25
Asio otus, see Owl, Long-eared
Aspens, 64, 65, 68, 77, 81, 84, 89, 123, 124, 135
Atchafalaya River (La.), 14
Atchafalaya Swamp (La.), 92, 94
Anticosti Island (Gulf of St. Lawrence), 24
Atlantic Ocean, 22
Audubon, J. J., 102
Auffenberg, W., 21
Aylward, D., 85

Bagg, A. C., 105
Baie Johan Beetz (Que.), 25
Bailey, W. J., *see* Liscinsky and Bailey
Baird, J. and C. Robbins, 145
Balsam, 124
Bancroft ,T. A., *see* Kozicky, Bancroft and Horneyer
Banding, recoveries, 150; repeats, 88; returns, 111–112; spring, 7–9; summer, 9–10; winter, 10–11
Bannerman, D. A., 161
Barnes, W., 100
Barnstable County (Mass.), 26–27
Barrows, W. B., 102
Barry, W., 170
Bartlett, L. M., *see* Sheldon, Shaw and Bartlett
Bartramia longicauda, see Plover, Upland
Becket, L., 108
Bedstraw, 79
Belleview, Fla., 21
Bennett, L. J. and P. F. English, 105, 123
Bent, A. C., 68
Berkshire County (Mass.), 27
Bermuda, 24
Beule, J. D., *see* Norris, Beule and Studholme
Beyer, O. E., A. Allison and H. H. Kopman, 70
Bill, *see* Skull
Birches, 63, 64, 68, 77, 81, 86, 89, 123, 124, 125
Bitterweed, 94

INDEX

Blackberries, 26, 64, 66, 67, 86, 94, 123, 124, 134
Blankenship, L. H., 30, 40, 42, 54, 55, 57, 62, 64, 65, 68, 70, 76, 85, 102, 125, 126, 133, 141, 154, 155
Blueberries, 26, 64, 66, 67, 86, 94, 123, 124, 134
Bluestems, 63, 94
Bobcats, 152
Bobwhite (*Colinus virginianus*), 79, 140
Bogardus, A. H., 17
Bonasa umbellus, *see* Grouse, Ruffed
Bond, R. S., *see* MacConnell and Bond
Boston (Mass.), 27
Bowles, D. G., 34
Boyer, G. F., 25
Brain, position of, *see* Skull
Branta canadensis, *see* Goose, Canada
Brewster, W., 46, 51
British Museum of Natural History, 156
Brodkorb, P., 21
Broods, behavior of chicks, 73–75, of hen, 72, 73, 74, 75–77; cover, 77; duration as entity, 74–75; feeding, 74; flying age, 73, 75; growth of chicks, 73; hatching dates, 72; method of hatching, 72–73; number of, 72
Brown, C. P., 102, 105
Brown, J. L., 60
Bubo virginianus, *see* Owl, Great Horned
Bump, G., R. W. Darrow, F. C. Edminster, and W. F. Crissey, 48
Bursa of Fabricius, 38
Butler, A. W., 102
Byron, P., 70

Canton (Mass.), 27
Cape Cod (Mass.), 26, 27, 58, 96, 153
Cape May (N.J.), 19, 96, 136, 147
Capella gallinago, *see* Snipe, Wilson's
Captive woodcock, 82
Capture methods, decoy trap, 7, 8, 177–180; draw nets (Europe), 10; funnel trap, 8, 10; hand netting, 10, 11, 75; head light, 11; Japanese mist net, 8, 10, 85, 181–183; tape recording, 7; use of dogs, 8–9; victrolas, 8
Catgut, 94
Cedars, 63, 64
Census methods, counts of singing males, 19, 46, 54, 55, 56, 57, 127–130; direct counts on wintering grounds, 19; flushing rate, 130
Charadrius semipalmatus, *see* Plover, Semipalmated
Cherries, 63, 123
Chestnut, 86
Chicks, *see* Broods
Chocolate plant, 93
Cinquefoil, 79
Clark, E., 10
Clark, W. O., 90
Clarke, C. H. D., 24, 25, 122
Classification, 31
Clay, T., 156
Cloquet Eperimental Forest (Minn.), 65
Cobb, S., 32, 33, 36
Cocklebur, 93
Colinus virginianus, *see* Bobwhite
Colorado, 24
Columba livia, *see* Pigeon, Domestic
Connecticut, 70, 145
Connecticut River, 12, 96
Corn fields, 126
Cottam, C., 79
Courtship, altitude of flights, 44, 54; behavior on ground, 44, 46, 49–51; copulation, 46, 47, 49–53; description of flight, 44, 46; duration, dawn, 56, evening, 55–56, seasonal, 56–59; effect of weather, 54–55 ff.; flights, fall, 58, summer, 58, winter, 58–59; grounds, females *peenting* on, 52–53, male replacement on, 47–49; latitude, 56; light intensity, 55; notes, 44, 51–53; peaks, 54, 57–58; variation, in individual birds, 55, between birds, 54
Cowardin, L., 54
Crabapple, 63, 69, 123, 124
Crissey, W. F., *see* Bump, Darrow, Edminster, and Crissey
Crows, 152
Cumberland, F., 36
Cumberland Neck, Nova Scotia, 22
Curlews, 78
Cypress, 14

Dalapon, 133
Dangler, E. W. and W. H. Marshall, 63, 64, 102

Darrow, R. W., *see* Bump, Darrow, Edminster, and Crissey
Davis, F. W., 52
Dawson, W. L., 102
D.D.T., 157
Deer, White-tailed, 18
Delaware, 58, 96
Dewberry, 123
Dewitt, J. B., 157; *see also* Stickel, Dodge, Sheldon, Dewitt, and Stickel
Dexter, R. W., 79
Dieldrin, 158
Dodge, W. E., *see* Stickel, Dodge, Sheldon, Dewitt, and Stickel
Dogwoods, 63, 79, 85, 93, 123, 124, 135
Dowitcher, Short-billed (*Limnodromus griseus*), 32
Drainage, 18, 94, 138, 139, 153
Drummond Island (Mich.), 124
Duck, Wood (*Aix sponsa*), 18
Duke, G. E., 54, 55
Dukes County (Mass.), 26–27
Duvall, A. J., 37, 86

Ears, 36
Eaton, W. H., 102, 105
Edminster, F. C., 140; *see also* Bump, Darrow, Edminster, and Crissey
Eggs, color, 68; hatching dates, 70–71; number, 69–70, 71; size, 69
Elaphe obsoleta, see Snake, Pilot Black
Elder, 79
Elizabeth Islands (Mass.), 27, 96
Eneminger, A. B., 90
England, 81
English, P. F., *see* Bennett and English
Ereunetes pusillus, see Sandpiper, Semipalmated
Erithacus rubicula, see Robin Redbreast
Essex County (Mass.), 17, 85
Evergreens, 123
Eyes, 35–36, 62
Eynon, A. E., *see* Leopold and Eynon

Farner, D. S., 145
Fayle, D. C. F., 81
Feeding habits, 82–94; adaptations, 34, 78; amount required, 80; fall, 79, 89–90; food, 79–84, animal, 79, vegetable, 79; method of feeding, 34, 35; summer, 79, 85–89, 90; winter, 79, 90–94
Ferns, 79, 86, 123, 126
Finland, 144
Fisher, A. K., 18
Fitzpatrick, W. A., 58
Fleabane, 93
Florida, 21, 22, 24, 26
Flycatcher, Pied (*Muscicapa hypoleuca*), 144
Forbush, H. A., 26
Forester, Frank, *see* Henry Herbert
Fort Albany (Ont.), 24
Foscue, N. J. and C. L. White, 98
Foxtail, 93
Franklin County (Mass.), 27
Frog, 79

Gates, G. E., 80
Geis, S. A., *see* Locke, Stickel and Geis
Georgia, 22, 70, 96
Gilfillan, W. C., 100, 102, 105
Glascow, L. L., 11, 14, 19, 25, 36, 42, 59, 80, 82, 84, 90, 91, 92, 93–94, 95, 96, 101, 103, 107, 109, 112, 131, 132, 138, 139, 141, 142, 143, 149
Godwits, 78
Goldenrod, 123
Goodrum, P., *see* Reid and Goodrum
Goose, Canada (*Branta canadensis*), 127
Goose, Gray Lag (*Anser anser*), 35
Gordon, S., 168
Goudy, W. H., 54, 55
Grant, W. C., 80
Grasses, 63, 64, 93, 94
Greeley, F., 30, 38, 42, 86, 141; *see also* Sheldon, Greeley and Kupa
Greenbrier, 124, 126
Greenbush (Me.), 57
Gregory, J. U., 43
Grinnell, G. B., 76
Griscom, L., 105
Grosse Tete (La.), 92
Grouse, Ruffed (*Bonasa umbellus*), 10, 39, 48, 60, 69, 72, 76, 117
Gull, Herring (*Larus argentatus*), 32, 60
Gum trees, 14, 93
Gunn, W. W. H., 7

Habitat, 64, 85, 93, 94, 128, 129, age, 124–125, succession of vegetation, 123; basic requirement, 122; cover

INDEX

Habitat (*continued*)
density, 125; fall, 122–126, beaver ponds, 138, soil, 122, vegetation, 62–66 ff., 122–126; Massachusetts, 27–28; shade, 85, 122; signs of use, 124; spring, 62–66, 77, 192; summer, 85–88; winter, 92–94.
Hackberry, 93
Hagar, J. A., 54
Haile (Fla.), 21
Hampden County (Mass.), 27
Hampshire County (Mass.), 27
Hancock County (Me.), 83
Handley, W. R. C., 81
Hardhack, 89, 124, 126
Harris, B. K., 30
Harris, S. W., 21
Hatching, *see* Eggs, Broods
Hawes, A. F., *see* Hawley and Hawes
Hawk, Cooper's (*Accipiter cooperii*), 152
Hawk, Sharp-shinned (*Accipiter striatus*), 152
Hawley, R. C. and A. F. Hawes, 64
Hawthorne, 85, 123, 124
Hayne, D. W., *see* Stickel, Hayne and Stickel
Heath, 26, 64
Hazelnut, 123
Hens carrying chicks, 76–77; *see also* European Woodcock
Heptachlor, 157
Herbert, Henry (Frank Forester), 16, 17, 19
Herman, W. J., 81
Heteralocha, *see* Huia
Hickey, J. J., 142
History, current populations, 19–20; earliest record, 16; hunting, 16–19; laws, 18–19; numbers before Civil War, 16; population trends since Civil War, 17–20
Hochbaum, H. A., 48
Hoffman, J., 164
Home range, 28
Homing, 110–113, to breeding grounds, adult female, 111, adult male, 110, juvenile female, 112, juvenile male, 111; to wintering grounds, sex ratio of returns in Louisiana, 112–113
Horneyer, P. G., *see* Kozicky, Bancroft and Horneyer
Howard, Ethan, Jr., 49
Howell, A. H., 26
Hudson Bay, 24
Huia (*Heteralocha*), 78
Hunting, after Civil War, 17; before Civil War, 16; dogs, 118–119; "dusking," 89; early regulations, 18; early twentieth century, 18; eighteenth century, 16; estimated number shot in North America, 121; guns, gauge, 117, shot size, 117; market, 18, 19, 85, 99–100; nineteenth century, 17; success and flushing rates, 119, 120

Iberville (La.), 14, 26
Illinois, 24
Indiana, 24
Injuries, 149
Iowa, 22

Jacksonville (Fla.), 26
James Bay (Ont.), 24, 25, 80
Johnson, B., 160
Jourdain, The Rev. F. C. R., *see* Witherby, Jourdain, Ticehurst, and Tucker
Juniper, 63, 64, 66

Kansas, 22
Keewatin (Can.), 24
Keniston, A., 101
Kentucky, 56, 70
King, E. P., 108
Kiwi, New Zealand, 34
Knight, J. A., 126
Kopman, H. H., *see* Beyer, Allison and Kopman
Kozicky, E. L., T. A. Bancroft and P. G. Horneyer, 128, 129
Kupa, J., *see* Sheldon, Greeley and Kupa

Lack, D., 47, 144, 170
Lake Abitibi (Ont.), 25
Lake Huron, 48
Lake Nipigon (Ont.), 25
Lake Superior, 25
Lanyon, W. E., 21, 141
Lapwing, European (*Vanellus vanellus*), 37, 145, 167
Larus argentatus, *see* Gull, Herring
Latham, R. M., 58
Laws, *see* History

Le Blanc, J. F., 108
Le Jeune, P., 16
Leopold, A. and A. E. Eynon, 55
Lespedeza, 94
Lewis, E. J., 17
Lewis, E. L., 170
Ligon, J. D., 21
Limnodromus griseus, see Dowitcher, Short-billed
Lincoln, F. C., 70
Liscinsky, S. A., 8, 20, 25, 58, 76, 80, 82, 85, 99, 102, 105, 119, 123, 124, 126, 133, 134, 137, 138, 153; and W. J. Bailey, 10
Locke, L. N., W. H. Stickel and S. A. Geis, 155
Locust trees, 81, 93, 96, 123
Long Island (N.Y.), 27, 96
Louisiana, 10, 11; study areas, 14–15; 18, 22, 24, 25, 26, 43, 58–59, 70–81, 90, 92, 93, 94, 96, 130, 131, 138, 139, 143, 148, 149, 151, 152, 158; parishes, 14, 26
Lowery, G. H., Jr., 109
Lumsden, A., 48
Lynch, J. J., 43

MacConnell, W. P. and R. S. Bond, 134
Maine, 10, 11, 12, 22, 40, 47, 52, 65, 67, 69, 70, 71, 79, 80, 81, 83, 89, 96, 122, 124, 141, 148, 151, 154
Management, bag limits and hunting season, 136–137; census, 127, 132; creation of habitat, burning, 93, 134, 139, cutting, 133, grazing, 126, 136, land management practices, 138–139, planting, 134, spraying, 133–134
Manitoba (Can.), 22
Manitoulin Island (Ont.), 48
Maples, 63, 77, 123
Marshall, W. H., 42, 64, 65; *see also* Dangler and Marshall
Martha's Vineyard (Mass.), 66, 96, 153
Martin, F. W., 10, 11, 38, 52, 86, 89, 129, 138, 141, 143, 144, 145, 159
Maryland, 56, 148, 149
Massachusetts, 7, 17, 26–28, 40, 41, 44, 46, 52, 96, 123, 125, 128, 129, 134, 136–137, 140, 148, 149, 152, 155, 157
Massachusetts Division of Forests and Parks, 12
Mathis, J., 79
Mayfield, H. K., 63
Mayer, J., 10
Mayr, E., 60
McIlhenny, E. A., 11, 43, 112, 151
Meadowsweet, 63, 124
Melospiza melodia, see Sparrow, Song
Mendall, H. L. and C. M. Aldous, 8, 22, 25, 26, 28, 30, 43, 46, 52, 58, 62, 65, 67, 68, 69, 70, 74, 76, 77, 81, 83, 85, 96, 102, 104, 108, 112, 122, 136, 141, 143, 148, 150, 151, 152, 153, 154, 167
Merovka, L. J., 10, 109, 112
Metropolitan Water Commission, 12
Michigan, 10, 19, 20, 22, 40, 42, 52, 54, 57, 64, 65, 68, 70, 80, 88, 89, 120, 124, 129, 130, 133, 134, 138, 141, 144, 154, 155
Migrations, altitude, 108; arrival in wintering grounds, 109–110; by state: Ala. 98, Ark. 100, Del. 96, 98, Fla. 98, Ga. 96, 98, Ill. 99, 100, 102, 104, Ind. 99, 100, 102, 104, Ia. 104, Ky. 99, 101, 105, La. 96, 98, 100, 101, 104, 105, 107, 109, 110, Me. 96, 98, 102, 104, 106, 107, 108, Md. 101, Mass. 96, 98, 99, 101, 103, 104, 105, 107, 108, 109, Mich. 99, 100, 102, 104, 105, 106, Minn. 100, 102, 104, Miss. 99, 102, Mo. 100, 101, N.H. 96, 105, 106, 107, N.J. 96, 98, 99, 101, 104, 106, N.Y. 98, 102, 105, 106, N.C. 98, 104, O. 99, 100, 102, 104, 105, Pa. 98, 99, 102, 105, 106, S.C. 96, 98, Tenn. 98, 99, 105, Tex. 100, Vt. 107, Va. 98, 99, 102, Wis. 100, 104, 106; by province: N.B. 96, 103, 106, 107, N.S. 96, Ont. 99, Que. 99; departure of "native" birds, 108–109; fall, 43, "flights," early fall, 96, flocking, 108; late fall, 96; rate, 103–104; routes, 95, 96–100 ff., Atlantic, 96–99, Central, 99–100, Western, 100; see aggregation, 42; spring, 101–104, departure from La., 101, to Mass., 102–103; effect of weather, 103, 107–108
Mikula, E., 19
Miller, D. R., 79, 82
Millet, 93
Minnesota, 19, 20, 22, 42, 63, 64, 66, 117, 148
Mississippi, 22, 24
Mississippi River, 14, 24, 94

Missouri, 22, 24
Montana, 24
Monogamy, 46
Montgomery, J. P., 14
Monticola sp., 35
Moon, effect of, 11, 54–55
Moosehorn National Wildlife Refuge (Me.), 132
Mortality, accidents, 152; chemical spraying, 156–158; diseases and parasites, 154–156; hunter kill, 150; nest destruction and abandonment, 153; predation, 152–153, bobcats, 152, Cooper's Hawks, 152–153, crows, 152, feral dogs, 153, Great Horned Owls, 152, house cats, 153, Long-eared Owls, 153, Pilot Black Snakes, 153, Sharpshinned Hawks, 152, shrikes, 153, skunks, 152, weasels, 152; weather, 150–152
Mouseley, H., 10, 31, 32, 36, 76, 160, 161
Murchie, W. R., 80
Muscicapa hypoleuca, see Flycatcher, Pied
Musselman, T. E., 105

Nantucket Island (Mass.), 26, 27, 64, 124
National Audubon Society, 136
Natural Chimneys (Va.), 21
Naushon Island (Mass.), 27
Nebraska, 22
Neponset River (Mass.), 27
Nesting, behavior of incubating hen, 66–67, 71–72; cover, 67–68; distance from singing grounds, 68; ecology, 67; characteristics of nest, 67; success, 68–70; *see also* Eggs, Broods
New Brunswick (Can.), 8, 20, 22, 42, 56, 71, 96, 119, 123, 136, 141, 157
Newburyport (Mass.), 57
Newfoundland, 22
New Hampshire, 96, 141
Neuhaus, W., 34
New Jersey, 19, 22, 24, 70, 79, 96, 137, 141, 145
New Mexico, 21
New York, 17, 20, 22, 141
Nice, M. M., 47, 60, 153
Norris, R. T., 10, 26, 57, 59, 91, 109; *see also* Norris, Beule and Studholme
Norris, R. T., J. D. Beule and A. F. Studholme, 7, 46, 50, 52, 65
North Carolina, 22, 24, 56
North Dakota, 22, 96
Nova Scotia (Can.), 22, 42, 79, 117, 119, 148
Nutting, W., behavior of birds on summer field, 39, 87–88

Oaks, 14, 26, 64, 65, 86, 96, 123
Oberholser, H. C., 26
Ohio, 22, 24
Oklahoma, 22, 24
Ontario (Can.), 22, 24, 122
Owl, Great Horned (*Bubo virginianus*), 143, 152
Owl, Long-eared (*Asio otus*), 152

Parrots, 37
Parus caerulens, see Titmouse, Blue
Patuxent Wildlife Research Center, 155
Pay, L. M., 165
Payne, F., 52
Peabody, G., 17
Pelletier, A., 53
Pennsylvania, 7, 8, 10, 20, 22, 58, 65, 79, 82, 85, 117, 119, 123, 124, 126, 133, 134, 137, 141
Pettingill, O. S., Jr., 16, 18, 26, 30, 38, 46, 50, 52, 58, 68, 69, 70, 71, 73, 79, 81, 85, 96, 104, 152, 153, 167
Phasianus colchicus, see Pheasant, Ring-necked
Pheasant, Ring-necked (*Phasianus colchicus*), 60, 140
Phillips, J. C., 17
Phoenicurus phoenicurus, see Redstart
Pigeon, Domestic (*Columba livia*), 35
Pigeon grass, 79
Pines, 12, 14, 26, 63, 64, 65, 66, 84, 94, 96, 126, 135
Pitelka, F. A., 50, 99, 102
Plover, Golden (*Pluvialis dominica*), 167
Plover, New Zealand Wrybill (*Anahynchus*), 78
Plover, Semipalmated (*Charadrius semipalmatus*), 32
Plover, Upland (*Bartramia longicauda*), 50, 61
Plumage, coloration, 29–30, 38, protective, 29, 38, 66, 74; molt, 37, 72

Pluvialis dominica, see Plover, Golden
Pointe Coupee (La.), 14, 26
Polygamy, 46–47
Population dynamics, breeding age, maximum, 143, minimum, 143; density, 147; factors effecting wing collection, 144–145; life equation, 147; longevity, 143–144; sex ratios, 141, discrepancy in and possible reasons for, 141–142; sex and age ratios, 38, from fall wing collection, 75, 140–141, analysis of, 145–147; survival, 146
Populations, composition of netted summer birds, 86; *see also* History, Census
Preble, E. A., 24, 25, 76
Prince Edward Island (Can.), 22
Puchalski, W., 164

Quabbin Reservation, description of, 12–14; 66, 85, 128, 136, 148
Quebec (Can.), 16, 22, 24, 81

Rabbit, cottontail (*Sylvilagus* sp.), 37
Ragweed, 79, 93
Range, ancient, 21; current, 21–23, northern limits of breeding range, 22, 23, 24–25, southern limits, 22, 23, 25–26; in Massachusetts, 26–28; winter, 24, 26
Raspberries, 79, 123
Ravenswood (La.), 92
Reardon, J. D., 133
Redcross (La.), 92
Redshanks, 167
Redstart, European (*Phoenicurus phoenicurus*), 144
Refuges, 136
Reid, V. and P. Goodrum, 25, 42, 130, 131
Research, future needs, 173–175
Rhode Island, 20, 22
Rice, J. H., Jr., 43
Ripley, T. H., 58, 140
Roberts, T. S., 102
Robbins, C., *see* Baird and Robbins
Robin Redbreast (*Erithacus rubicula*), 35, 144
Rose, 93
Rowan, W., 34, 35
Russell, D. M., 70, 101

St. Andrew's Cross, 94
St. Landry (La.), 14
St. Lawrence River (Que.), 25
Salamanders, 25
Sandpipers, 61
Sandpiper, Pectoral (*Erolia melanotos*), 61
Sandpiper, Semipalmated (*Ereunetes pusillus*), 32
Sandpiper, Stilt (*Micropalama himantopus*), 61
Sandys, E. and T. S. Van Dyke, 18
Saskatchewan, 24, Black River, 24, Athabasca Lake, 24
Schack, K. *et al.*, 170, 171
Schenk, J., 169
Schorger, A. W., 76
Scolopax rusticola, see Woodcock, European
Scott, I. G., 19, 101, 106
Sedges, 79, 94
Selby, P. J., 168
Senses, 34–37, hearing, 36–37; sight, 35–37; smell, 34; tactile, 34; taste, 34
Sesbania, 93
Sex, methods of external separation of male and female, 30–31, bill length, 30, outer primary width, 30, size, 30
Shakespeare, W., 160
Shaw, F. R., *see* Sheldon, Shaw and Bartlett
Sheldon, W. G., F. Greeley and J. Kupa, 38
Sheldon, W. G., F. R. Shaw and L. M. Bartlett, 156
Sheldon, W. G., *see also* Sheldon, Greeley and Kupa; Sheldon, Shaw and Bartlett; Stickel, Dodge, Sheldon, Dewitt, and Stickel; Stickel, Sheldon and Stickel
Shutesbury (Mass.), 7
Singing grounds, 86, distance to diurnal cover of male, 28, 65; height of surrounding trees, 63; longevity of, 66–67; origin of, 65–66; shape, 62–63; size, 62–63; soil, 65; topography, 63; vegetation, 63–65, 66, 192
Skull, 31, 32, 36, bill, 33–35; position of brain, 32–33
Skunk, 153
Slater, H. H., 162
Smartweeds, 79, 93

Smith, E. A., 120
Smith, F., 80
Smith, R., 53
Smith, W., 43
Snake, Pilot Black (*Elaphe obsoleta*), 153
Snipe, Wilson's (*Capella gallinago*), 11, 46
Soils, 14
South Carolina, 24, 96, 151
South Dakota, 22
Sperry, C. C., 79
Sparrow, Song (*Melospiza melodia*), 154
Spruces, 26, 77, 124, 135
Stebbings, J. H., 80, 81
Stegeman, L. C., 81
Steinfatt, O., 72, 163, 166, 167
Stewart, R. E. and J. Aldrich, 47, 148, 149
Stickel, L. F., *see* Stickel (W. H.), Hayne, and Stickel; Stickel, Dodge, Sheldon, Dewitt, and Stickel
Stickel, W. H., D. W. Hayne and L. F. Stickel, 158
Stickel, W. H., W. E. Dodge, W. G. Sheldon, J. B. Dewitt, and L. F. Stickel, 158
Stickel, W. H., *see also* Locke, Stickel and Geis
Studholme, A. T., *see* Norris, Beule and Studholme
Suffolk County (Mass.), 27
Sumacs, 94
Sumpweed, 93
Sylvilagus sp., *see* Cottontail rabbit

Tampa (Fla.), 26
Tennessee, 24
Territoriality, 60–62, behavior of males, 61, of females, 61; definition, for woodcocks, 60–62
Texas, 22, 24, 26, 84
Thompson, D. Q., 126
Thorp, T. B., 109
Thrush, 35
Ticehurst, N. F., *see* Witherby, Jourdain, Ticehurst, and Tucker
Tie vine, 93
Tinbergen, N., 60
Titmouse, Blue (*Parus caerulens*), 35
Tongue, 35
Tree of heaven, 81
Tuck, L., 46
Tucker, B. W., *see* Witherby, Jourdain, Ticehurst, and Tucker
Tufts, R. W., 30, 42
Turkey, Wild (*Meleagris gallopavo*), 60
Turner, F., 21
Twilight flights, 44–46, behavior, 44–46; effect of weather on, 88; fall, 89; summer, 85–89; winter, 90–92; *see also* Weather
Twitle, R. A., 101
Tyrell, J. B., 24

U.S. Bureau of Sport Fisheries and Wildlife, 12, 127, 156

Van Dyke, T. S., *see* Sandys and Van Dyke
Vanellus vanellus, see Lapwing, European
Van Pelt, A., 25
Vermont, 12, 29, 42, 84, 89, 112, 123, 125, 126
Violet, 79
Virginia, 24, 70
Voice, 38–40, 51–53

Wayne, A. T., 43, 150, 151
Weasels, 152
Weather, as factor in twilight flights, 54–55; as mortality factor, 19; effect on feeding behavior, 131; effect on migration, 103, 107–108; effect on winter captures; *see also* Twilight flights
Weed, G., 19
Weeden, R. B., 68, 125
Weight, 40–43, fall, differences by age and sex, 42; female, 42; male, 42; spring, 40–42; summer, 41, 42; winter, 42 43
Wentworth, B. C., 35
Westfall, C., 40, 52, 57, 103
West Virginia, 22, 24
Wetherbee, D. K., 61, 69, 73
Wetmore, A., 21
Whippoorwill, 9
White, C. L. and N. J. Foscue, 98
White Water Game Refuge (Minn.), 66
Willow, 89

Wings, method of separation by sex and age, 37–38, 203, 204; shape, 37; twitter sound, 40
Winsk River (Ont.), 24
Wisconsin, 20, 22, 38, 42, 141
Witherby, H. F., F. C. R. Jourdain, N. F. Ticehurst, and B. W. Tucker, 161, 165, 168
Wolfson, A., 43
Woodcock, European, 10, 31, 34, 72, 76, 160–172; breeding habits, number of broods, 72, 165, copulation, 164, courtship flights, 163–164, duration of breeding, seasonal, 164–165, hen carrying chicks, 76, 167, incubation by hen, 166, nest and eggs, 165–166; food, 168; habitat, 168; history, 160; hunting regulations, 170–171; migration, 169; physical differences from American Woodcock, 31, 161–162; populations, 169–170; range, breeding, 162–163, winter, 163
Worcester County (Mass.), 27
Worms, cover types, 81; ecological distribution, 81, 82; geographical distribution, 80–81; origin in America, 80–81; soil types, 81
Wright, B. S., 8, 24, 56, 71, 76, 103, 119, 123, 125, 157

Yankee weed, 93
Yarmouth (N.S.), 22
York Factory (Can.), 24